A SENSE OF SPACE

A SENSE OF SPACE

A LOCAL'S GUIDE
to a FLAT EARTH, *the*
EDGE *of the* COSMOS,
and OTHER
CURIOUS PLACES

JOHN EDWARD HUTH

The University of Chicago Press * Chicago and London

The University of Chicago Press, Chicago 60637
The University of Chicago Press, Ltd., London
© 2025 by John Edward Huth
Published 2025
Printed in the United States of America

34 33 32 31 30 29 28 27 26 25   1 2 3 4 5

ISBN-13: 978-0-226-84442-8 (cloth)
ISBN-13: 978-0-226-84443-5 (ebook)
DOI: https://doi.org/10.7208/chicago/9780226844435.001.0001

Library of Congress Cataloging-in-Publication Data

Names: Huth, John Edward, author.
Title: A sense of space : a local's guide to a flat Earth, the edge of the cosmos,
    and other curious places / John Edward Huth.
Description: Chicago : The University of Chicago Press, 2025. | Includes
    bibliographical references and index.
Identifiers: LCCN 2025009578 | ISBN 9780226844428 (cloth) | ISBN
    9780226844435 (ebook)
Subjects: LCSH: Space perception. | Space astronomy. | Cosmology.
Classification: LCC BF469 .H88 2025 | DDC 153.7/52—dc23/eng/20250324
LC record available at https://lccn.loc.gov/2025009578

♾ This paper meets the requirements of ANSI/NISO Z39.48-1992
(Permanence of Paper).

*

TO KAREN BERYL AGNEW

*

CONTENTS

PREFACE

The stars close to the horizon beckoned. I couldn't keep my eyes off them because I expected sunrise soon from the orientation of Cassiopeia and the Big Dipper. Silhouetted against this backdrop, the mast of the traditional outrigger *Jitdam Kapeel* swung to and fro in the heavy sea. On the chase vessel, most of us were seasick. A reporter on board quietly chummed her stomach contents over the boat's railing.

This was in 2015 when I had joined an expedition to the Marshall Islands to understand the tradition of navigation using ocean waves. Alson Kelen, who captained the outrigger, was our navigator, and we followed.

Traditional navigation in the Marshall Islands uses ocean waves as the primary reference, and often stars, when visible. According to the wave pilots, there are always swells present from the east, west, north, and south. The waves interact with the islands and atolls, giving clues to the outrigger's location. They can reflect off land or refract around an island creating a pattern of crossing waves in the lee side of the atoll. Sometimes land blocks the waves completely.

Alson navigated on the outrigger without instruments, using methods he had learned from his elders in the Marshall Islands. Onboard the chase vessel, I tracked our progress with a GPS receiver and nautical chart. I also recorded the ambient conditions: wind speed and direction and waves I could sense.

At three in the morning we radioed Alson and asked him if he could estimate his location. He replied that he thought we were fifteen miles to the southeast of Aur, our destination. We checked the GPS location and chart and noted that Alson was spot on. I was astonished how he could know precisely where we were after hours of darkness in the bouncy sea.

I was participating in the voyage as a physicist. Along with a specialist in computer simulation of waves, we brought our toolkit of Western science to understand how the navigators in the Marshalls pulled off their remarkable feats of wayfinding. It was not a one-way street. One of Alson's mentors, a navigator by the name of Captain Korent, was seeking validation of the Marshallese tradition of navigation from Western scientists as a means of cultural revival.

Well, the Sun did rise, and we ducked in between two atolls into calmer waters, and eventually into the lagoon of Aur, where we landed at the village of Tabal. We spent several days in the village, and I tried to absorb as much of the local culture as I could. Although we can find Western culture on the Marshall Islands, the inhabitants retained a strong sense of pride in their ancient traditions of navigation. There is the ubiquitous presence of stick charts—wood lashed together to act as both a teaching aid for apprentice navigators and a map that indicates the kinds of waves you'd experience at different locations. Folklore includes tales of how navigation came to the islands, which were passed down from one generation to the next. In one origin story, a woman named Litarmelu was taught navigation by three alien men who arrived at her atoll in a ship shaped like a box. Over time, Litarmelu became skilled in wave piloting and taught the techniques to her sons.

I learned that the navigators of the Marshall Islands are practitioners of a science. While it may appear different from what I may have thought of as science in the West, it has an empirical framework on which to hang observations. The traditions of navigation extended from the perceptions of waves and stars into the folklore and cultural traditions. There is no clear dividing line.

My conjecture is that the navigators are exploiting regularities in the geography of the Marshall Islands. The islands and atolls of the Marshall Islands consist of two chains running roughly northwest to southeast. The trade winds are the prevailing winds and blow from east to west and produce distinctive wave patterns along the axes connecting any pair of atolls.

If this is the case, then the topography of the islands largely enabled the culture of navigation, which then extended into other aspects of the islanders' lives. Directions in Marshallese aren't given in terms of "left" or "right," which are body designators. Instead, they refer to a broader kind of directionality of place, like saying "the hut is on the north side of the path, nearest the ocean."

After several days on Tabal it was time for our return voyage. The trip back to the capital of Majuro was considerably calmer than the outgoing. So calm, in fact, that the wind died out in the middle of the night. With the luxury of having a GPS, I could see that the ocean currents were sweeping us off course to the west. Without the wind, the *Jitdam Kapeel* was at the mercy of the current. Soon after sunrise the wind began to pick up again, and Alson had the canoe sailing close-hauled to regain the track to Majuro. At that point we couldn't see any sign of land in the distance, just ocean. This wasn't surprising; the atolls of the Marshall Islands are low-lying and you can't spot land unless you're practically right on top of it. Because of the current, we might miss Majuro altogether. Then at about eleven in the morning, we finally spotted Majuro in the distance. As we got closer we could make out landmarks and eventually spied the buoy marking the entrance to the lagoon.

Reflecting on the voyage, and other work with anthropologists, I began to wonder how the models of space we deal with in our science have shaped the way we think in the West. There is an emerging view among psychologists that the same parts of our brain involved in spatial maps are also tasked with creating our social map: who we are close to, and where we fit in a social hierarchy. More generally, it seems that as our science imagines space in new ways, these new images are reflected in our social understandings.

When I returned from the Marshall Islands, I began a new voyage that ended in the book before you, reimagining the evolution of concepts of space, from the time of Egyptians onward. I went from a flat Earth with mystical explanations for motions in the heavens, to more mechanistic explanations by Greek astronomers and astrologers, all the way to our current understanding of the universe, and the unanswered questions.

As you'll see in the pages to come, Western science expanded our horizons of space to progressively larger and smaller scales over the centuries. Although our minds have evolved to comprehend space on terrestrial distances, we're capable of extending this cognition to cosmological and subatomic scales, albeit where new metaphors are necessary. In addition to the domain of the large and the small, there are also new concepts of space that emerge when we strive for a unified theory, one of Einstein's quests.

One theme of the book is whether a perspective of space is referenced to people. If I were to give you directions, I might say, "Travel a mile ahead to the first traffic light and then turn right." On the other hand, I might say, "Travel west a mile to the first traffic light and then turn north." Depending on who you are, you may or may not find either set of instructions terribly helpful. As you'll see, these two kinds of representations of space are quite common: people-centered and people-free. It affects how our mind "sees" our immediate surroundings—either referenced to us, as in "left," "right," "front," "back," on one hand. A people-free space might use "north," "south," "east," "west" as more general directions. We're capable of holding both perspectives and often will translate between the two.

Marshall Island stick charts certainly exhibit these two perspectives. In these lattices of sticks, patterns of the sticks represent various ocean features found near islands and in between. In some cases the representations are independent of people with currents near islands or a path of disturbed water connecting pairs of islands indicated. There are other interpretations of stick charts that put the outrigger canoe in the center and show wave systems from the perspective canoe's crew members.

In the development of concepts of space in the West, there's a similar schism between representations that are centered on people and those that are independent. In astrological charts the person is centered in space, surrounded by the planets, Sun, Moon, and stars. Their lives are represented by the alignment of the celestial objects at the moment of their birth. On the other hand, the modern version of the solar system places the Sun at the center with the Earth orbiting along with the other planets. No humans are required.

The space imagined by Galileo and Newton also did not require humans, and it did not have a center but was more of a scaffolding or stage on which the laws of motion played out. One important assumption was a sense of absolute time: All clocks tick at the same rate.

Even in the modern era, there are social connections to new visions of space. Einstein challenged the notion of absolute time in his special theory of relativity.

The centrality of humans persists to this day among some theoretical physicists. The state of the art of theory predicts an almost uncountable number of different possible universes, all with different constants of nature. Some physicists believe that the only way to explain our universe is that it is one of the rare ones that allows for human life.

As I move through different Western visions of space in the following chapters, it's useful to keep in mind that science is a human endeavor like the Indigenous tradition of voyaging in the Marshall Islands. We have diagrams that act as teaching and analysis tools, much like the stick charts. Scientists are navigators, often striking out with new experiments into the unknown. But navigators and scientists don't exist in a vacuum; we are embedded in culture, often inspired by, and create new metaphors in the ongoing voyage of discovery.

Here I've assembled a travelogue through time and space for an ever-widening awareness of our horizons and collective imagination. Navigating the waters in the pages that follow, I hope to show how our social perceptions of space necessarily impact how we do our science. In turn, when science reveals new senses of space—from the realm of quantum mechanics to the origin of the cosmos—I argue that it changes our social perceptions. Like the waves and currents flowing among the atolls, the interplay signs a way forward.

MENTAL MAPS *of* SPACE, MEMORY, *and* SOCIETY

"Where are you going with this?"
  "I'm not following you"
  "You seem so distant"
  "That's all behind us now"
  "My mind is wandering"
  We often use the language of navigation through physical space to describe how we move through time, with the past "behind" us, and the future "ahead." It is as if our motion in time is a stroll. Likewise, many of us use language of motion and position in social contexts. Physical proximity stands in for intimacy. We say that a couple is close or have split apart. Or engagement, "You seem distant." Social status is associated with height: "getting in on the ground floor" or "I have to speak with my higher-ups."
  Physical navigation—getting from one place to another—is fundamental to life. Iron-fixing bacteria will swim in the direction of the Earth's magnetic field. Arctic terns will migrate from the Arctic to the Antarctic and back again every year. Humans have managed to explore and inhabit nearly every niche of the planet. Not surprisingly we all have the inbuilt smarts to purposefully find our way. But physical migration is only one part of what it means to be human. We live in cultures that have structures and people that we must also navigate if we're to live a purposeful life.

We navigate websites, we escape from traps laid. We generalize the concept of finding our way to all sorts of situations. If someone is behaving erratically, we say that they've lost their bearings or slipped their moorings. We speak of following a moral compass.

Life is a journey and much of our imagination also occupies a kind of mental space, searching for a path. Stories, dreams, movies, poetry, and songs exist in real or imagined landscapes. When we speak of "being between a rock and a hard place," we are invoking Homer's voyage of Ulysses and the passage between Scylla, the cliff-dwelling monster, and the whirlpool Charybdis.

If you're aware of the parallel metaphors of spatial and social language, it sometimes becomes difficult to hold a conversation of any length of time (see?) without invoking the language of navigation. When my students and I began to discuss spatial metaphors in a seminar, it became extremely halting as every accidental use of a spatial indicator caused laughter. We can't get anywhere without using this language.

In *Metaphors We Live By*, by psychologists George Lakoff and Mark Johnson, a chapter is devoted to the spatial/social metaphors in the English language. My work with anthropologists demonstrated to me that this was quite common in other languages as well. There is an increasing belief among some cognitive psychologists that the parts of the brain associated with spatial reasoning are also engaged in social reasoning. This parallelism may be a hint as to why spatial expressions are used widely in a social context. To understand the possible parallelism and gain some perspectives, I first turn to how neuroscientists and cognitive psychologists believe our minds organize space.

While there are ways we perceive space internally, there is also the actual physical space the brain occupies in our cranium. The brain has different regions that serve specific purposes. For example, the auditory cortex on the side of the brain near the ear analyzes sound. There are longer connections of neurons that link these many specialized regions together into what we might perceive as the core of our being or consciousness. Specifically, the seat of our mental map lies in a part of the brain called the limbic system. Some call this region the "reptilian brain" because it is far older, in an evolutionary sense, than the outer neocortex found in mammals.

The limbic system has two neighboring regions, the hippocampus and the entorhinal cortex, both of which are implicated in spatial cognition. These are indicated in figure 1.1. Much of what we do know comes from experiments on animals where neuroscientists implant electrodes in their brains and then monitor individual cells that light up when the animal engages in certain behaviors. In addition to this, we can learn from humans who suffer damage to parts of their brains and infer some functionality.

For our internal map: in the hippocampus there are cells called "place cells," neurons that are activated when a person or animal is in a specific location in an environment that they are familiar with. In the neighboring entorhinal cortex there are "grid cells" that are activated when the individual is in multiple locations in an environment that are regularly spaced, like the tiling on a floor or the backsplash of a sink. Both the hippocampus and the entorhinal cortex have dense interconnections of neurons linking the two together. As of now, we

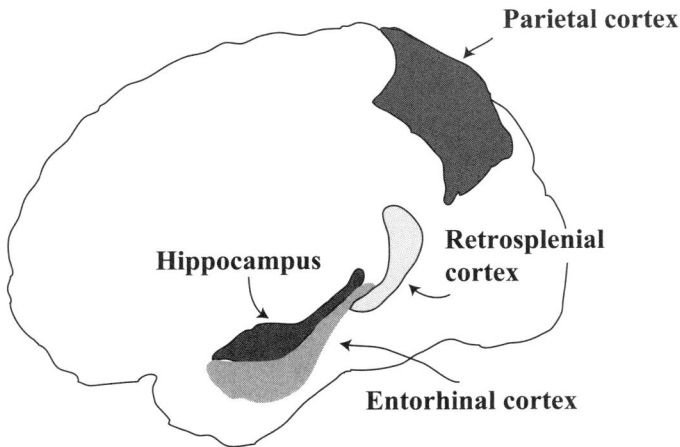

FIGURE 1.1 Parts of the human brain implicated in navigation.

4 * CHAPTER ONE

do not know precisely how this combination of grid and place cells act to create a mental map. This is an intense ongoing study and there are many conjectures.

What does the mental representation of space look like? Psychologist Edward Tolman asked this question in a 1948 paper, where he speculated that there are two possible forms of a mental spatial map. One kind is called "route knowledge," where space is represented as a series of possible paths, and the navigator knows only the paths and their interconnections. Another kind of mental map is called "survey knowledge," where space is laid out as if you are hovering over a landscape and can see all the possible two-dimensional interconnections at once.

Figure 1.2 shows a map of a portion of Lower Manhattan. The organization of streets in much of Manhattan is on a grid system of avenues running approximately north–south, and numbered streets running approximately east–west. The grid breaks down in Lower Manhattan and the streets begin to look higgledy-piggledy and more challenging to navigate.

How can you tell the difference between route and survey knowledge in the behavior of an individual? Short-cutting!! That is to say, finding a new path that may not be on your "list" of available paths. Say you're a tourist on your own looking at the sites in Lower Manhattan. You start out with a selfie next to the Charging Bull statue, but then want to visit the 9/11 Memorial, some 10 minutes' walk away. Armed with a plastic-coated map, or slavishly following your cellphone directions, you move approximately in the direction of the Memorial. First up Broadway, and then take a left onto Cedar. We would call this route knowledge (fig. 1.3).

Now, suppose instead of going solo, you have a friend familiar with Manhattan accompany you on your sightseeing tour. You two pose for a photo next to the Charging Bull and then begin the walk toward the 9/11 Memorial. Walking down Broadway, your friend sees that the grounds for the Trinity Church are open and suggests cutting across to slice a minute off the walk, and thus take in another sight

⭐ **9/11 Memorial & Museum**

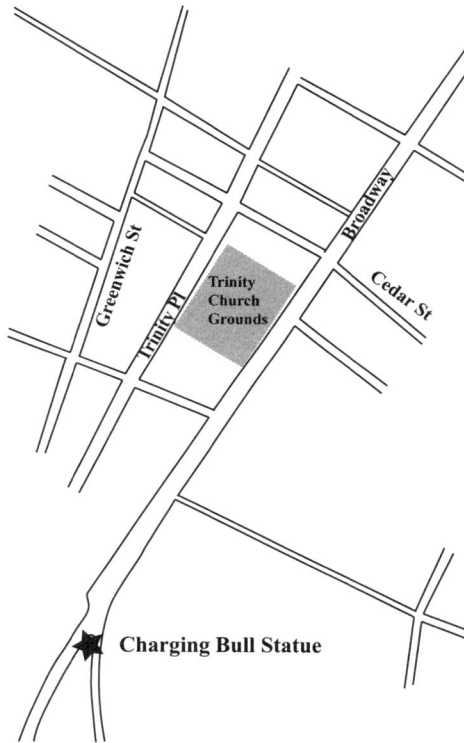

FIGURE 1.2 A section of Lower Manhattan with
three landmarks: the Charging Bull statue, the Trinity Church
and grounds, and the 9/11 Memorial and Museum.

in Lower Manhattan. We would call this survey knowledge, where
your friend is aware of the shortcut (fig. 1.4).

Now, back to Edward Tolman's experiments with route and survey
knowledge. He tested the possibility of route versus survey knowledge
in a classic rats-in-a-maze experiment. Tolman first trained rats to
navigate from an opening to a food source in a maze. The rats learned
the path to the food source. Then Tolman changed the maze in a way
that allowed a more direct pathway to the food. There were many

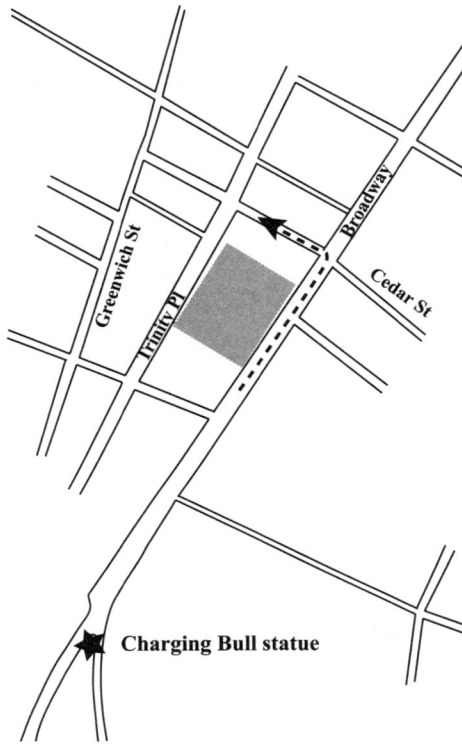

FIGURE 1.3 Part of a path someone with route knowledge might choose on the way from the Charging Bull statue to the 9/11 Memorial.

possible paths presented, including one that mimicked the original. The rats chose a more direct path, indicating that they possessed survey knowledge of their environment.

We have multiple cognitive maps depending on our environment. Think back to the area you originally grew up in, or the layout of your primary school. Depending on the circumstances, your mind can call up the map for the envisioned environment. These multiple maps can lead to interesting situations. One of these I call the waking-up-in-a-strange-bed-in-the-dark problem. Suppose you're traveling

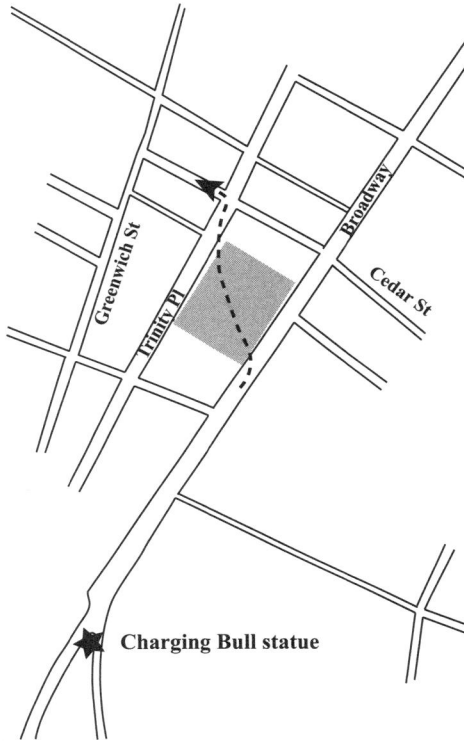

FIGURE 1.4 Part of a path someone with survey knowledge might choose on the way from the Charging Bull statue to the 9/11 Memorial. There is a shortcut through the grounds of the Trinity Church.

and are staying in a hotel in a faraway city. You wake from a dream, which, itself, has its own imaginary landscape. In the dark your first instinct is to believe that you're in your home bed, but something isn't right. You reach for the light switch or some familiar object on the bedside, but it's not where it's normally located. There's a moment of confusion and then everything falls into place, and you think, "Oh yes, I'm in X-and-such location, and I'm on the other side of the bed from where I normally sleep."

I had the strange experience of witnessing my brain "trying on" different cognitive maps for a good fit. To some extent, I attribute this to a nighttime cold remedy that gave me strange dreams. I had dreams within dreams within dreams and would wake up from one dream and think I was back into reality, only to discover I was in another dream. Each dream bore its own landscape. Then, in the morning, while I was half-asleep, half-awake, I heard the familiar pattern of the sounds of my daughter opening her bedroom door, walking down the hallway, and entering and closing the bathroom door. From this sequence, I "knew" I was now inhabiting reality and not another dream, but where was that? In a strange state of mindfulness, I could witness my brain as it tried out different cognitive maps. First, it tried on my old bedroom from when I was a ten-year-old boy. No, that wasn't a good fit. Then it tried on my house in suburban Chicago when I worked at an accelerator laboratory. No, that wasn't a good fit either. Then it tried on my house in Newton, Massachusetts, and suddenly the room spun around, and everything fit perfectly. The map stabilized.

The hippocampus is also implicated in a number of mental activities beyond cognitive maps. One of the most noteworthy of these functions is called episodic memory. This is when the individual retrieves a memory as an act of will. If, for example, I ask you to name your favorite high school teacher, you go into the files of your mind to retrieve the name, and likely the face of that teacher. Sometimes this takes time, and you might respond by saying, "I can see their face but forget their name off the top of my head."

It may not seem obvious that memories and a mental image of an environment would reside in the same part of the brain, but they both share things in common: relations and a sense of proximity. Memories come in categories, like the names of teachers you may have had or long-lost loves. You might "stroll down memory lane" or say that you've lost your "train of thought." There is an emergent view among cognitive psychologists that episodic memories are intrinsically visual in nature and, effectively, the brain is retasking the navigational parts that include knowing landmarks and their relative positions as a means of organizing memories.

MEMORY PALACES

The parallel between memories and spatial maps is not something new. It's been around at least as far back in time as the Roman states-man Cicero. In his work on rhetoric, *De Oratore*, Cicero writes of a mnemonic technique that's often referred to as a "memory palace." Cicero recounts the story of a poet, Simonides, who attended the ban-quet of a wealthy nobleman. In the middle of the festivities Simonides was called outside of the palace to receive a message. While he was outside, the roof of the building collapsed, killing everyone inside, but sparing Simonides. Many of the bodies were crushed beyond recognition. Simonides came to the aid of the grieving relatives to identify the bodies from his memory of where they were seated in the banquet hall. By realizing this association of people with places where the guests were seated, Cicero credits Simonides with the concept of a memory palace. In this concept, the mental image of the physical layout of a structure becomes a kind of scaffolding on which to hang and organize memories.

In order to give a long speech without notes, the orator must mem-orize a huge number of facts and then find a way of invoking them throughout. Utilizing the memory palace technique can be useful. It involves imagining a large building with rooms, each of which has its own unique furnishings, and passageways and stairs connecting the floors and rooms. Memories are then associated with specific rooms, and the speaker will imagine an exploration of the memory palace as the speech unfolds. The technique is effective and is still used to this day with many self-help books on improving memory.

The memory palace technique does not have to be limited to a palace or a house per se. The map in your mind could be a landscape or even imaginary. For example, Johannes Romberch, a German Dominican, wrote about an imaginary memory landscape invoked by Dante's structure of Paradise, Purgatory, and Hell in *Congestorium artificiose memorie* in 1520. Hell, for example, consists of a series of concentric circular ledges, each one populated by souls guilty of a particular sin. The second circle is for lust, the third for gluttony, and so forth. Paradise consists of a set of nesting spheres, each one associated with a virtue. The third sphere of Venus is for Lovers, the fourth sphere of the Sun is for The Wise.

TIME IN THE BRAIN

In addition to the cognitive map and episodic memories, the hippocampus is implicated in both future planning and imagining. Future planning, in a way, makes sense because this is just the temporal equivalent of navigating through an environment. The hippocampus also appears to be involved in the perception of time. Researcher Howard Eichenbaum found that "time cells" in the hippocampus fire when a particular time in an understood sequence of events occur and provide the basis for the temporal organization for episodic memories.[1] For example, you're given a tour of an art museum, and you see paintings by famous impressionists, then surrealists, and then cubists. You later recount the visit to a friend and the order of the painting schools on display: the time cells help organize the sequence of the memories. These were found in studies of epileptic patients who were about to undergo surgery and were asked to first memorize a sequence of words, and later recite the sequence while their brains were monitored with electrodes. A specific time cell would light up at the moment a specific part of the sequence was articulated. People with lesions impairing the function of their hippocampi could recall certain events but not piece them together in the order in which they occurred.

The time cells don't tick in a regular fashion like a stopwatch, but rather speed up or slow down, depending on the circumstance, or even emotion. This thought was voiced in a poem by Henry Van Dyke, originally composed as an inscription for a sundial:

Time Is

Too Slow for those who Wait
Too Swift for those who Fear,
Too Long for those who Grieve
Too Short for those who Rejoice;
But for those who Love,
Time is not.

HENRY VAN DYKE, *Music and Other Poems*, 1904

The sequencing of time cells for memories suggests the physical basis for the language of our experience as beings traveling through time. If I'm giving a lecture and want to turn to a new topic, I might say, "If you've managed to follow me up until now, I'll now turn to a related subject." This makes the lecture sound like I'm leading the students on a hike and taking a fork in the trail.

PERSONAL AND GLOBAL REPRESENTATIONS OF SPACE

Returning to the brain's map of physical space, in addition to the concept of route knowledge and survey knowledge, there is a question of whether the map has the individual at the center or whether the map is independent of the person. These two representations are respectively called egocentric and allocentric. There are multiple definitions for these terms. Neuroscientist Dr. Maria Kozhevnikov offers one take. She says egocentric representations locate "objects in space relative to the body axes of the self (left-right, front-back, up-down)," while the allocentric representation "encodes information of an object or its parts with respect to other objects. The location of one object is defined relative to the location of other objects."[2]

There are many other possible definitions, but they roughly are along these lines, although one could readily substitute "landmarks" for "objects." The term "landmark" might imply a greater degree of

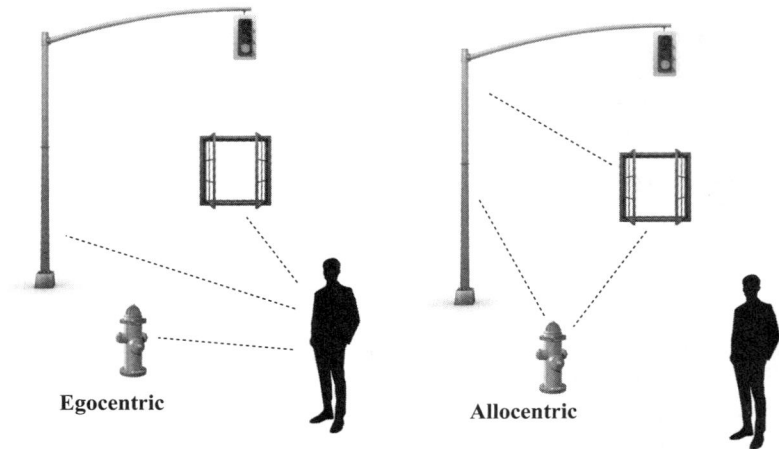

FIGURE 1.5 Egocentric and allocentric perspectives of
an individual with respect to landmarks.

permanence. A moving bicycle is certainly an object, but its location relative to the observer and more permanent features, such as a fire hydrant, is constantly changing. Figure 1.5 illustrates the two possibilities with relative position of landmarks referenced to the individual as an egocentric perspective and the relative positions of landmarks referenced to each other as an allocentric perspective.

What are the neural mechanisms for allocentric and egocentric representations? Cognitive psychologist Eleanor Maguire studied navigation and the brain. She held that humans, indeed, have different regions of the brain that are active in these two representations. Based on her work she suggested that the hippocampus is fundamentally allocentric, and another part of the brain at the back of the skull, the parietal cortex, is where the egocentric representations reside. According to Maguire the hippocampus holds the allocentric master maps, and these get translated into an egocentric representation for the parietal cortex to use in the exigencies of immediate navigation.[3]

As hinted in the preface, the difference between egocentric and allocentric perspectives can readily be imagined (fig. 1.6). Say you were

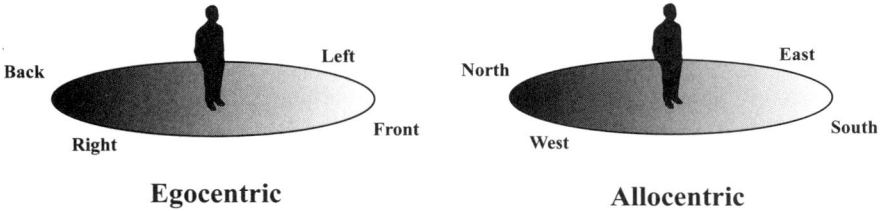

Egocentric   **Allocentric**

FIGURE 1.6 A more expansive view of the distinction between egocentric and allocentric frames of reference. The egocentric can refer to directions with respect to the body (left-right-front-back). The allocentric could refer to the cardinal points of the compass (north-south-east-west).

to approach me in Harvard Square and ask for directions. Thinking egocentrically, I might say something like, "Continue down Massachusetts Avenue until you reach the coffee shop, and then make a left." On the other, allocentric, hand, if I said, "Continue north on Massachusetts Avenue until you reach the coffee shop, and then turn west," most visitors might find this less than helpful. An alternative definition of the terms would be to suggest that an egocentric representation is referenced to front-back-left-right of an individual, and an allocentric representation might involve using east-west-north-south as designators.

The roads near Harvard Square can be confusing, but other locales may lend themselves to a more allocentric description. For example, the streets of mid-town Manhattan and Chicago are laid out on grids, so saying, "Drive east on 34th Street and then turn north onto FDR drive," would be well understood to a New Yorker.

One curious phenomenon is a variation in the language of orientation across different cultures. While most Westerners prefer directions in terms of the egocentric left, right, forward, and backward, there are languages that are more allocentric. Austronesian languages like Polynesian and Micronesian variants tend to avoid body-based direction indicators. In some cases, it's more geographically based. For example, "my home is down the path and the last house on the lagoon side." This is not to say that Westerners are purely egocentric or Austronesian speakers are mostly allocentric. In English we might

talk about "going inland" or "seaward," which has allocentric conno-
tations. Or a Polynesian speaker might refer to the side of a canoe
with an outrigger as being one side or the other.

Cognitive psychologist Lera Boroditsky reported that in Porm-
puraaw, a small Aboriginal community in Australia, the inhabitants
use the cardinal points north-south-east-west as direction indicators.
So, you might say, "pass the water pitcher to your north," if you were
at the table and speaking in their language, Kuuk Thaayorre.[4]

Throughout the book I will keep returning to representations of
space that shift between egocentric and allocentric. For example, Aristotle's model of the universe had a stationary Earth at the center and
a series of spheres surrounding the earth representing the Moon, Sun,
planets, and the firmament of stars, all rotating at different speeds.
On the other hand, the Copernican model of the universe placed
the Sun at the center, with the planets orbiting around. One could
extend the concept to suggest that Aristotle's model was egocentric,
while the Copernican model was allocentric.

As an analysis tool in future chapters, I've broadened the definitions to consider a spatial representation egocentric if it necessarily
involves humans. This could be in the case where the Earth is the
center of the universe. Another example would be if a representation
necessarily involves an observer as an integral part. On the other hand,
a representation could be considered allocentric if it does not rely on
humans. An allocentric perspective would also be represented by a
substantial jump in broadening our understanding of space, including
the horizons. The horizons can be to the large, as in the limits of the
observable universe, or to the small, as in the case of explorations at
high-energy particle accelerators.

I don't want, however, to convey the impression that allocentric and egocentric representations are a perfect binary, nor an all-
encompassing explanation of how we perceive space. It seems that we
all have the capabilities of holding both perspectives. Nonetheless,
it becomes a useful implement in the toolkit to examine emerging
concepts of space.

SOCIAL MAPS

Like Simonides at his banquet, it's not difficult to imagine a simple extension of spatial maps to include individuals and their relation to you. It can be as straightforward as placing neighbors in their respective houses in your neighborhood. As long as an individual is situated in a fixed physical location with respect to you, their home or office becomes as much of a landmark as a church spire. This pairing of physical place to the individual can break down when they appear in an uncommon location. For example, if you bump into your dentist at the gym, there can be some initial confusion with a thought, "They look familiar, but I can't quite place them."

There is an interesting question of how an individual can become a kind of landmark independent of physical location. Clearly, their home can be a landmark, but there's a question of how the actual individual can be identified. Their faces are something we can identify, and we think "ah-ha," this is the person who is my supervisor, colleague, brother, teacher, and so on, understanding their relation to us. How might this identification work?

Let's consider the case of the discovery of a "Jennifer Aniston neuron" identified by neuroscientist Itzhak Fried and colleagues, where they found a specific neuron that began firing when a subject was shown photographs of the actress.[5] In other subjects, person-specific neurons fired when shown photographs of other celebrities, like Bill Clinton. Now, the actual people-neurons are certainly not acting alone but are highly interrelated to other neurons that aid in the recognition of faces, identity, and relation that culminates in the thought, "Ah, this is Jennifer."

Beyond the physical location of an individual, there is also our relation to the individual. Recognizing photos of Jennifer Anniston, we would think, "I see her in the movies and on TV, I know who she is, but she certainly doesn't know who I am." But, in the case of coworkers, family, neighbors, or friends, there is a kind of connection between the us and our relations. The relationship could be a recognition of "this is my student," or "this is my co-worker," or "this is my spouse," and so forth.

An interesting clue in socialization comes from individuals who have suffered hippocampal injuries where they can recognize a face but cannot place them in the context of a social setting—not knowing, for example, what their role is in an organization or their relation to the individual. This is similar to the affliction of people with lesions in their hippocampus who cannot recall a sequence of memories.

The relation among individuals may also be represented as a kind of map, but in a different variety of space. Here the metaphors of physical proximity associated with degree of interaction (they're close, they're distant) and of height with status (I must speak with my higher-ups), may be telling.

Neuroscientists Matthew Schafer and Daniela Schiller from Mount Sinai Hospital in New York have advanced the hypothesis that the hippocampus plays a central role in organizing social networks into a kind of two-dimensional space where relationships are mapped.[6] The dimensions of the social map are proximity of affiliation in one direction and the power status in the other direction. This could be in an organization where the power structure is a hierarchy of CEO, directors, managers, administrators from top to bottom. Affiliation can be within a family with spouse being closest, siblings a bit farther, cousins, uncles and so forth more distant still (fig. 1.7).

This also raises the question about whether the concept of egocentric spatial representations spill over into egocentric social relations. An organizational chart may capture the allocentric representation of individuals. This might be with the big boss at the top of the chart, their administrators underneath, and then below that it branches out to the workers. This works at the abstract level, but to the person working in the office, their view may be a situation of having to cope with a perfectionist boss, and the structure seems more focused on the immediate relationship.

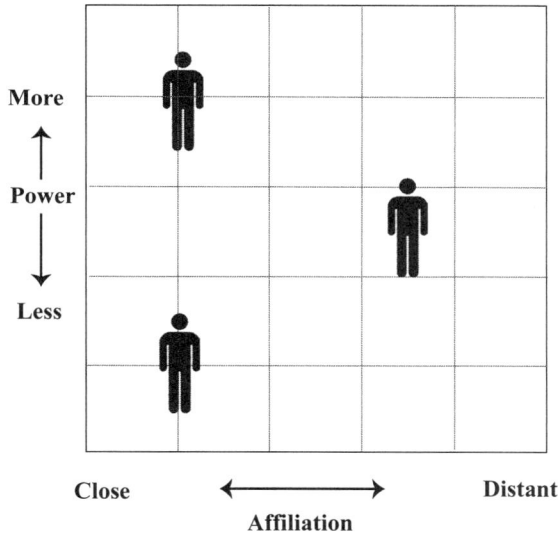

FIGURE 1.7 Matthew Schafer and Daniela Schiller's suggestion of a possible two-dimensional map of social relations based on the affiliation on one hand and power on the other.

The permanence of landmarks for a physical cognitive map is an issue. Perhaps a landmark is torn down, or the patterns of one-way streets gets shifted. This can be temporarily disorientating and takes some time to effectively create an edited map of the surroundings. If the suggestion of a social map is correct, then we might consider individuals as the landmarks that guide us as we navigate social situations. For example, a minister might be consulted to resolve a dispute between two parishioners.

This spatial-social analogy may be helpful in grasping the process of grieving. Our social landscape is predicated on a kind of timelessness of the "landmarks" of the ones closest to us. But when we lose a person, we lose one of our crucial landmarks and feel lost. A new place for the memory of the lost one must be established and space for a new view of close ones must be remapped. In some cases there is a feeling that the loved one is on a long trip, or perhaps arrived at a physical destination like heaven.

It is remarkable that this one part of the human brain, the hippocampus, seems to be implicated in so many scenarios, all of which have the underlying theme as different kinds of map:

Cognitive maps
Memory
Future planning
Imagining
Time sequencing
Social relations

This being the case, it is no small wonder that our cultures produce metaphors where social structure is mimicked in spatial terms. The powerful get the penthouse apartments and the not-so-powerful get the basements. In Samoa the powerful have houses closest to the center of the common village green and the not-so-powerful have houses nearer the periphery, close to the outhouses.

But what about our concepts of space itself? Prior to contact with the West, the Polynesians were able to successfully navigate thousands of miles across the Pacific, often using the stars as guides. During the Norse expansion in the North Atlantic from 900 AD onward, the Viking sailors used the Sun as a guide and even had a crude sense of latitude from the position of the Sun in the sky. These feats of navigation involved not only a kind of mental map of the vast swaths of ocean they traversed but also the peoples who lived in the far-off lands.

Then there are the spaces where our imaginations travel. Where do dead souls go in the afterlife? What are the geographies of heaven and hell? And what of places in the sky and under the Earth? Are they real? And, if they are real, what are their characteristics, and how do they relate to us as humans? Are they inhabited by intelligent creatures who have minds that work like ours?

The question of language designators as egocentric or allocentric, for example, has been studied by both cognitive psychologists and anthropologists. One aim of these studies is to see whether individuals in certain cultures have a better sense of direction if they are restricted to using the language of only allocentric place designators, rather than purely egocentric.

There is also a question of how concrete representations of space can merge into the mystical. We tend to think that this is not a property of Western cultures, at least not the practice of science, but for many cultures there is no distinction.

One of the anthropologists I worked with was Susan Montague.[7] She related her experience learning the language of Kaduwaga Village in the Trobriand Islands, located off the east coast of New Guinea. When she was living there, she was perplexed by the prefixes of "o" and "wa" that were used to designate places. They seemed to be confusingly mixed. The "o" and "wa" distinction arises from a Trobriand cosmography where unborn/dead souls inhabit a realm underneath the Earth—these are the "wa" beings. A "wa" being can rise into the womb of a pregnant woman and become a live human or an "o" being. How this applied as a prefix to place names mystified Montague at first.

She described how she might be walking down the "o" path or the "wa" path. The inhabitants would correct her when she made an error, but she couldn't understand what distinguished an "o-type-space" from a "wa-type-space." Over time, she learned that it was related to the Kaduwagan concept of space and reincarnation. "O" space is a space where people can congregate and effectively agree upon what is there concretely. "Wa" space, conversely, is more indeterminate and a space that is not as agreed upon.

The "o" space is the space of the living, occupied, established, embedded in the culture, like the inside of a general store, where people gather. The "wa" space is the space of the dead, unoccupied, not agreed upon, and somehow malleable and independent of culture, like a flat region hidden behind someone's house, untouched and ungardened.

In my attempts to understand the secrets of navigation across many cultures, I've come to understand that there are cultural elements that we would deem as empirical but that blend into what we in the West would call superstition. For example, the use of stars to orient or navigate is something Westerners might recognize as valid and utilitarian, but associating the same stars with ancestors of mythical heroes seems superstitious. Many tend to neatly divide beliefs into what they feel is a clean delineation of scientific versus unscientific. Yet I've come to see more of a continuum.

In the philosophy of science there is something appropriately called the "demarcation problem," where it is difficult to find a way to characterize the boundaries among what we might call serious science, pseudoscience, and pure imagination. In looking at the history of concepts of space from the ancient Egyptians to our modern understanding, there are countless demarcation problems where the human imagination may spawn new cultural manifestations. It seems impossible to create clear demarcations in the same way Montague found seamless connections between the Trobriand geography, cosmography, language, culture, and a view of the spiritual realm.

My colleague Gerald Holton, a historian of physics, reflected on these seamless cultural connections in the context of Einstein:

> The existence of both splendid scientific theories and splendid products of the humanistic imagination shows that despite all their other differences, they share the ability to build on concepts that, as Einstein put it, are initially of a "purely fictional character." And even their respective fundamentals, despite all their differences, can share a common origin. That is to say, at a given time the cultural pool contains a variety of themata and metaphors, and some of these have a plasticity and applicability far beyond their initial provenance.[8]

The interplay between spatial and social cognition has proven to be a powerful lens for studies in anthropology. Addressing the linkages of spatial and social metaphors across a range of cultures is a daunting task, however. On the other hand, the development of Western concepts of space from the ancient Greeks to the modern is rich in cultural connections and has a historical arc that allows us to see the interplay of these over the centuries.

At first blush it seems unlikely, but there is a surprising back-and-forth between concepts of space and culture in science. On the one hand, social and cultural factors can create inspiration for new visions of physical space. On the other, new visions of space, in turn, can spawn unique cultural manifestations that have a surprising longevity.

The tension between egocentric and allocentric perspectives emerges as a common theme. Naively, one might have thought that Western physics and astronomy to be purely allocentric. We think science becomes more universal over time. But, with an expansive definition of egocentric and allocentric, there are historical tendencies to adopt an egocentric perspective. This trend persists even now and has proven controversial.

I will not dwell exclusively on the egocentric/allocentric divide but just note that it appears in many guises. There are other paths in science to traverse using the social/spatial lens. A superstitious practice can lead to genuine breakthroughs. Sometimes scientific notions can devolve into magical thinking.

We are swimming in a sea of culture that both informs new visions of space and allows for new visions of what it means to be human.

FROM *a* FLAT EARTH
to EARLY COSMOLOGIES

My spouse and I travel annually to the Caribbean and, while there, we frequently visit a small roadside restaurant run by a devout woman named Sonia. On one of our trips in the early 1990s we learned that Sonia's adult daughter had died after being hit by a car while walking down the road in front of the restaurant. She was grieving and explained how she took solace. "I know that God's looking after her. Do you know how I know there's a God?" We were silent and she continued, "I know there's a God because the Sun sets over there." She gestured to the west. "And the Sun rises right over there every morning, without fail" and gestured toward the east. "The only way the Sun can possibly return each day is because of God."

How much of your knowledge is the result of direct experience? How much of your knowledge comes from external sources with trusted experts informing you? Maybe your knowledge comes from posts on social media. It's sometimes said that we live in a "post-truth era" where disinformation abounds. Many question the validity of "facts." There is the famous line about "alternative facts" regarding the 2017 presidential inauguration. It's worth pausing to examine the modern and ancient beliefs in a flat Earth as a window into knowledge, culture, and experience.

There is a recent uptick in the belief of a flat Earth. The *Boston Globe* reported the strange case of a New Hampshire man, Jason Torres, who in his desire to prove that the Earth was round ended up convincing himself that the Earth is flat.[1] According to the article, Torres spent a large amount of his time doing measurements in his quest. One attempt involved taking photographs of the town of Hull from four miles across Boston Harbor. He believed that the curvature of the Earth would make some of Hull disappear under the horizon but reported that he didn't see such any effect.

More recently, Georgia GOP district chair Kandiss Taylor hosted two flat-Earthers on her podcast, "Jesus, Guns, and Babies." In her interview with David Weiss ("Flat Earth Dave") and Matt Long, Taylor said, "The people that defend the globe don't know anything about the globe." Dave replied, "If they knew a tenth of what Matt and I know about the globe, they would be Flat-Earthers."[2]

Flat-Earthers hold that the world is a flat circular disk resembling the United Nations emblem, with the North Pole at the center and the farthest known territories a ring of mountains in what the rest of us call Antarctica (fig. 2.1). In their model the Sun circles overhead in a path that gives darkness to some areas and light to the others, depending on where it is in its orbit. Flat-Earthers maintain that photographs of the Earth from outer space are all doctored and there is a massive conspiracy to present it as a sphere. Only *they* know the truth.

Popularizers of science like Neil DeGrasse Tyson and Brian Cox fired back at the trend. In a social media rant Cox went off on the flat-Earthers: "There is absolutely no basis at all for thinking the world is flat. Nobody in human history, as far as I know, has thought the world was flat."[3]

Well, there have been many in human history who thought the world is flat. Let me ask the reader to introspect for a minute and ask yourself if you have any direct experience of the Earth as a sphere. I'm not talking about seeing photographs taken from the Moon, which the flat-Earthers claim are bogus; I'm talking about direct experience.

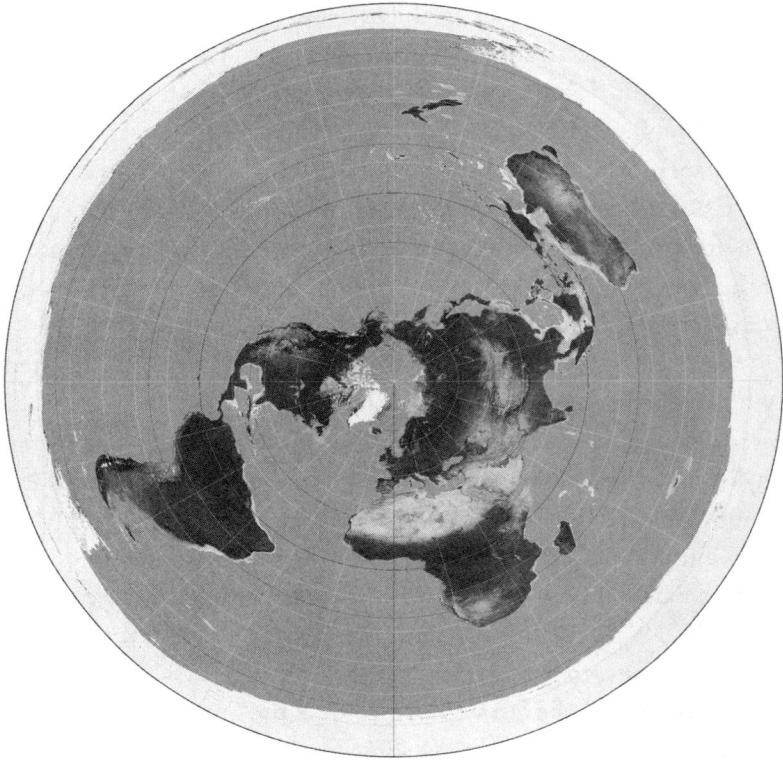

FIGURE 2.1 Modern flat-Earthers hold that the Earth is a flat
disk with the North Pole at the center. Credit: Daniel R. Strebe,
August 15, 2011, black-and-white rendering by the author.

Anthropologist Rick Feinberg studied the seafaring habits of the
sailors on the island of Anuta. In the final paragraph of his book on
the topic, he describes some discussion upon his imminent departure,
when he realized that the fishermen believed the Earth was flat:

> One day, several men came by to ask me where America was located.
> When queried as to what they meant, they asked if it were near the center
> of the world or closer to one of its edges. After a few moments of discus-
> sion, it became apparent that they viewed the world as flat (although they
> readily understood my explanation of the moving horizon in terms of
> the Earth being round and thought that this made a great deal of sense).[4]

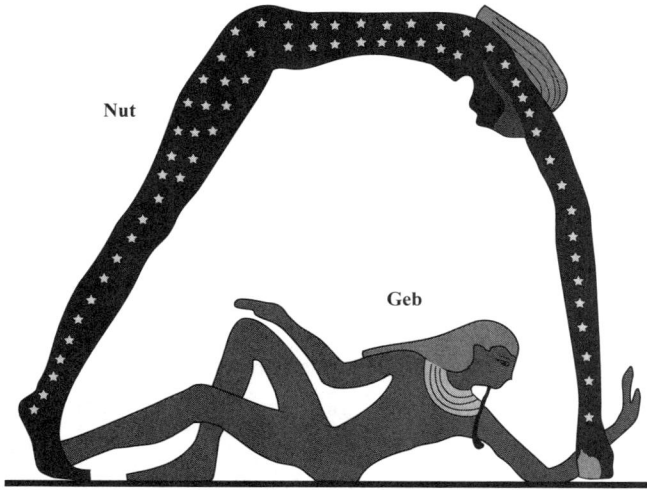

FIGURE 2.2 Nut and Geb.

It's not crazy to think of the Earth as flat, particularly if you don't travel very far. Some cultures never had a conception of the difference between a flat and a spherical Earth. Such a thought already requires some degree of abstraction: which of two (or more) geometric models describes the totality of the surface we inhabit? Without this kind of abstract thinking, we can't even contemplate the question.

There is a curious question for believers in a flat Earth: how does the Sun move from the west at sunset to the east at sunrise during the night? The modern flat-Earthers would say that the circular path of the Sun in the sky accomplishes this. What about the ancient Egyptians?

In ancient Egypt (approximately 3000–1600 BCE), there were multiple mythologies of the Sun and the sky. In one version the sky was viewed as a dome formed by the body of the goddess Nut, whose feet were planted in the east and whose head and arms extended to the west. Nut arched over Geb, the god of the Earth, who lay in recline (fig. 2.2). In one version of the mythology, the Sun journeys along the abdomen of Nut during the day, then she swallows the Sun in the evening, and she expels it from her hind quarters at dawn.

There is another myth where the Sun's return path to the east is through the Underworld. The idea that dead souls also inhabit this dark nether region has a certain logic. In one Egyptian variant, *Duat*, the underworld, is a region where the Sun (Ra) makes the return journey from west back to east. The setting Sun in the west is often associated with death and, not surprisingly, the idea that the dead inhabit the region where the Sun journeys back to the east at night might seem compelling. The navigation of dead souls in *Duat* was of considerable interest, and a kind of guidebook for the dead, often referred to as the Egyptian Book of the Dead, has been found in tombs. Our friend Sonia's belief that God is evidenced by the Sun's return to the east in the morning recalls the ancient Egyptian mythology of the Sun traversing the underworld populated by dead souls seems compelling.

Another effect that the ancient Egyptians surely noticed was the relative motion of the Sun against the background of stars, traversing all 360 degrees of the sky in one year. We nominally see this as a shift of the constellations that are overhead just after sunset. In winter, we see one set of constellations, and in the summer, another.

Egyptian priests would regularly scan the sky for "meanings," in an astrological sense. Their rituals of sky-gazing gave rise to the twenty-four-hour cycle of day/night. One annual event that had immense significance was the heliacal rising of Sirius. Unless you are well versed in archeoastronomy, you might be wondering: "What is a heliacal rising?"

Depending on your latitude and the time of the year, stars will appear in different parts of the sky just after sunset and just before dawn. In the mid-latitudes like Cairo, some stars will never set (e.g., Polaris), while others, like Sirius, will rise and set. The rising and setting stars always appear to move to the west relative to the Sun by about one degree per day. However, there is a patch of sky where a star will not be visible because it's too close to the Sun. At a certain time of the year, Sirius can be seen to set just after sunset. Then Sirius disappears for a few weeks while it's too close to the Sun to observe. After those weeks have passed, Sirius can be seen again, but now rising just before dawn. This rising before dawn is what is known as a *heliacal rising*.

Sirius is the brightest star in the sky, and the Egyptian priests noted that the heliacal rising of Sirius preceded the annual flooding of the Nile. Although it's a complete coincidence, the Egyptians viewed this occurrence with great reverence. Sirius is in the constellation *Canis major*, the Big Dog. It was thought that the brightness of Sirius combined with the heat of the Sun to create the hot months of July and August. This is how the hot summer months became known as "dog days." The heliacal risings of certain stars were often seen as harbingers of the seasons.

The further development of astronomy and astrology was a prominent feature of Babylonian culture from roughly 1000 BC onward. This was at a period where astronomy and astrology had no dividing line separating them and a major purpose of celestial observations was to predict events on the earthly domain. Babylonians made meticulous reports of observations of the positions of stars and planets, predicting their motions, and noting contemporaneous events on Earth for future reference. For example, they believed that flooding and famine follows a lunar eclipse. There is no known physical correlation between a lunar eclipse and flooding, but it's quite possible that historically there was a coincidence between the occurrence of an eclipse and a flood and this "stuck" in the historical record.

ASTRONOMY IN ANCIENT GREECE

Astronomy in Greece built on traditions that arose in Egypt and Mesopotamia. For example, the twenty-four hours in a day comes from the Egyptian priests. The ancient Greeks developed a succession of models that described the motions of celestial objects. The Sun-returning-to-the-east problem became less a mythological explanation and more a mechanical question. Perhaps the culmination of astronomy in ancient Greece is the creation of celestial almanacs and the means to calculate the positions of objects in the sky by Alexandrian Claudius Ptolemy (roughly 100–170 AD). His schema persisted until the seventeenth century when Johannes Kepler improved upon them with a modern model of the solar system.

The earliest mention of astronomy in Greece we have is from Homer in the *Odyssey* (eighth or ninth century BCE). In the *Odyssey* Homer describes how Ulysses uses stars as a navigational guide. In the *Iliad* Homer also describes the world as inscribed on a shield with the known lands surrounded by an encircling ocean. Hesiod (roughly 700 BCE), a poet, wrote of the time of harvest time coinciding with the heliacal rising of the Pleiades, the star cluster also known as the Seven Sisters. Herodotus, a historian, wrote of a solar eclipse occurring during a battle in the war between the Lydians and the Medes. He noted that the philosopher Thales of Miletus (600 BC) had predicted this eclipse the year before. Thales is regarded as the first of the pre-Socratic philosophers who broke with mythical explanations and offered up more science-like descriptions of phenomena.

We don't know how Thales was able to predict the eclipse, but the Babylonians had noted that solar eclipses only occurred during a new Moon. This was a strong clue that the Moon is closer to the Earth than the Sun. Also, the circular shape of the disk obscuring the Sun in an eclipse also gives a hint that the Moon is masking the Sun. Thales noted that the lunar phases can be explained by the Sun's illumination of its surface and viewed the Earth as being a flat disk floating in an encircling ocean.

ANAXIMANDER

Here we have our first mechanical explanation of how the Sun gets back to the east at dawn. Anaximander, a student of Thales, described a model where the Earth was a short cylinder floating freely in space (fig. 2.3). On the flat surface of the cylinder all the land of the world was surrounded by an encircling ocean. The Sun was a wheel in the sky that had a hole in it that opened to an outer region of fire. The hole in the rotating wheel gave the light of the Sun. The Moon likewise was on another rotating wheel. The axes of the wheels were inclined to the axis of the cylinder.

With the Earth portrayed as a cylinder floating free in space, the return path of the Sun back to the east in the morning gained a physical, as opposed to mythological, explanation. At night, the Sun passes "behind" the back of the cylinder of Earth. In Anaximander's model the stars were represented by a large hollow cylinder surrounding the wheel of the Sun. The passage of the Sun against the fixed background of stars throughout the seasons could be explained by the star-cylinder rotating at a different speed from the Sun. This gives rise to the relative path of the Sun through the stars. This model is noteworthy in that, while it provides an explanation of the motions of the Sun and stars, it still retains a flat surface of the Earth.

Following Anaximander, a series of flat-Earth systems were later proposed—most notably by Democritus (circa 400 BC), who is famous for his concept of atoms. We don't know much about the early philosophers and must rely on later writings, such as Aristotle, who tended to

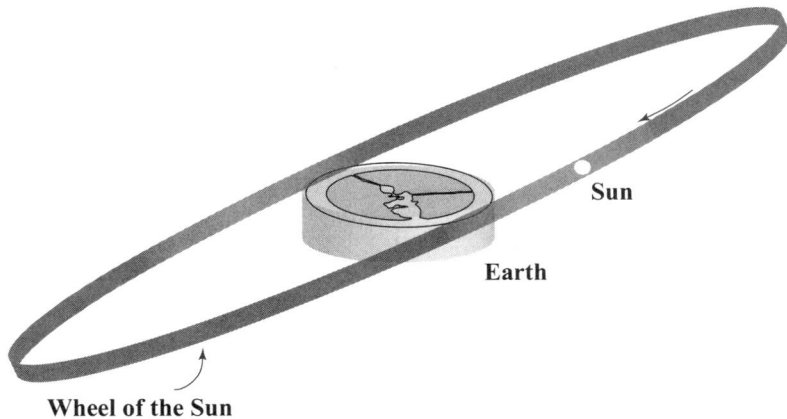

FIGURE 2.3 A depiction of part of Anaximander's universe with the Earth as a squat cylindrical disk and the Sun appearing as a hole in a wheel that faces fire.

be derisive of those who came before. Democritus also saw the Earth as flat, with a slightly upturned edge, supposedly to hold things in. He believed there were hemispheres of air above and below the disk, and then a sphere that held the Moon and Morning Star (Venus). Beyond this was the sphere of the Sun, the sphere of the planets, and finally the sphere of the stars. According to the historian Seneca, Democritus did not know the names or how many planets there were.

Early on, not much thought was given to the planets. They were confusing as they seemed to have random paths in the sky. To this day, few people can point out the planets in the sky. The paths of the planets keep them moving against the fixed background of stars, which is why they were named *planetes* or "wanderers." Their motion, along with the motions of the Sun and Moon, are restricted to a region of the sky called the *ecliptic*, which is perhaps a lucky accident. In the modern era we would say that this is a result of the solar system being confined to a plane. Despite the wandering the ancient Greek astronomers observed regularities in the planetary motions.

A SPHERICAL EARTH

A huge leap came from the Pythagoreans roughly around the time of Anaximander. These were followers of Pythagoras and their community originated in what is now Regio Calabria, Italy. While many important advances have been ascribed to them, the most significant for the purpose of this chapter is the notion that Earth is a sphere floating freely in space. We don't know precisely how the Pythagoreans arrived at this conclusion, but considering the modern flat-Earth conspiracy theorists there are multiple points listed below that are compelling reasons to believe in a spherical Earth.

PHILOSOPHY: A disk or truncated cylinder in Anaximander's model begs the question of what happens at the edge. An infinite plane is also challenging to contemplate. A sphere has no edge and is finite. If the sphere is large enough, it will locally appear like a plane and the spherical nature becomes more subtle.

RISING AND SETTING ANGLES OF CELESTIAL OBJECTS: Depending on your latitude, celestial objects will rise or set at different angles with respect to the horizon. Anaximander's Sun was on a tilted wheel and the inhabited surface was flat. A person anywhere on the cylinder's surface would see the same rising and setting angles. On the other hand, a person on a spherical Earth would see rising and setting angles that vary, depending on how far north or south of Greece you travel. This was noted by the Greek geographer Pytheas (350–306 BCE).

LENGTH OF DAY: Depending on the position of the Sun in the sky and latitude, the length of day will vary. In particular, the ancient Greeks labeled swaths of latitudes as climes, based on the length of day at the summer solstice. The further north (in the Northern Hemisphere), the longer the day at the solstice. This persists until past the Arctic Circle, where the Sun never sets at that time of year.

APPEARANCE OF STARS: Depending on your latitude, you'll see different stars in the sky. For example, the bright star Canopus is not visible from Boston, but when I vacation in Jamaica, it's readily visible. Likewise Canopus was not visible from Greece, but readily visible in Egypt. The Southern Cross is observable at about 15 degrees north latitude, but not in Greece.

APPEARANCE OF OBJECTS AT DISTANCES: As you sail away from a mountainous coastline, the bottom of the mountains disappears first, and then sequentially water appears to cover higher and higher parts of the mountains. The process reverses itself as you approach a mountainous coastline. As a caveat to this statement, the effect of light bending in the atmosphere can give strange results, so it is not foolproof.

ILLUMINATION OF CLOUD TOPS, MOUNTAIN TOPS AT SUNSET: Just after sunset or just before sunrise, the base of towering clouds may be in the shadow of the Earth, but the tops of the clouds are still illuminated.

LUNAR ECLIPSES: The shape of Earth's shadow during a lunar eclipse is clearly circular in shape. This may not rule out Anaximander's cylindrical model, but the different phases of the Moon demonstrate its sphericity. By extension, it would seem logical that the Earth and Sun are spheres.

I'm skeptical that the above arguments would convince a persistent flat-Earth conspiracy theorist, but the group presents a wonderful foil to understand how our views of space evolved. The ancient Greeks held that one half of the Earth was land occupied by people, and the other half was ocean. This was an outgrowth of the concept of an encircling ocean that was described by Homer in the *Iliad*, and Anaximander's squat cylinder. This notion persisted well into the period of Dante Alighieri, who built this into his model of the universe in the *Divine Comedy*.

PHILOLAUS AND OTHER PRE-ARISTOTELEAN ASTRONOMERS

Returning to one of this book's themes, we have the concept of *egocentric* and *allocentric* representations of space. For my purposes, I'm using a more expansive definition of these terms than the ones normally used in cognitive psychology. Recall that *egocentric* means a mental spatial map relative to the person and *allocentric* means a mental spatial map that exists independent of the person. In my more generous definition, an egocentric representation of space would have the Earth in the center, while in an allocentric model of space the Earth could be in motion. In more common usage, we typically call Earth-centered models *geocentric* and Sun-centered models *heliocentric*. As I further develop concepts of space, I will reuse *egocentric* and *allocentric* in other contexts beyond descriptions of the solar system.

We can contemplate whether Anaximander's model is one or the other in my more encompassing definitions. I argue that his construct is egocentric since all the motions associated with the sky happen away from the Earth. Although it's "floating" in space, the Earth is stationary and is the center of the wheels of the Sun and stars. These motions are described as observed by us on the ground.

Models that I would call allocentric do appear in ancient Greece. One notable image of space that I claim is more allocentric in its nature is attributed to the Pythagorean Philolaus from the fifth century BCE.

First, independent of any model, the observer on Earth sees two primary motions of the Sun and stars: (1) the Sun and stars appear to rotate around the Earth once every twenty-four hours; and (2) the Sun appears to move through the background of stars once a year. It was difficult to reconcile that the Earth was completely at rest and the entirety of the heavens moved with great speed around it, as in Anaximander's model. Philolaus created a model that had the stars fixed, and the apparent motions in the sky were the result of the motions of the Sun and Earth.

In Philolaus's model, the Sun and Earth both orbit a common center called the Central Fire (see figs. 2.4 and 2.5). The stars, on the other hand, were fixed and at a great distance. The Earth orbited the Central Fire once every twenty-four hours, and the Sun orbited once a year at a larger radius than the Earth These two motions give rise to the twenty-four-hour day-night cycle and explain the passage of the Sun against the background of fixed stars.

There are some additional wrinkles in Philolaus's model. For one, the occupied half of the Earth always faced away from the Central Fire, so it wasn't observed. The day-night sequence had to do with the Earth overtaking the Sun in their respective orbits around the Central Fire. Another feature of the model is a kind of counter-Earth that orbits the Central Fire opposite Earth called the *antichthon*. It's not clear what role the antichthon played. Perhaps it acted as a kind of counterbalance to the Earth. The occupied half of the Earth was turned away from the antichthon so, like the Central Fire, it wasn't visible.

Despite the peculiarities of Philolaus's model, it had several interesting features. It explained the motion of the Sun against the background of stars and it explained how the Sun gets back to the east at dawn. Perhaps most important, it was more of an allocentric

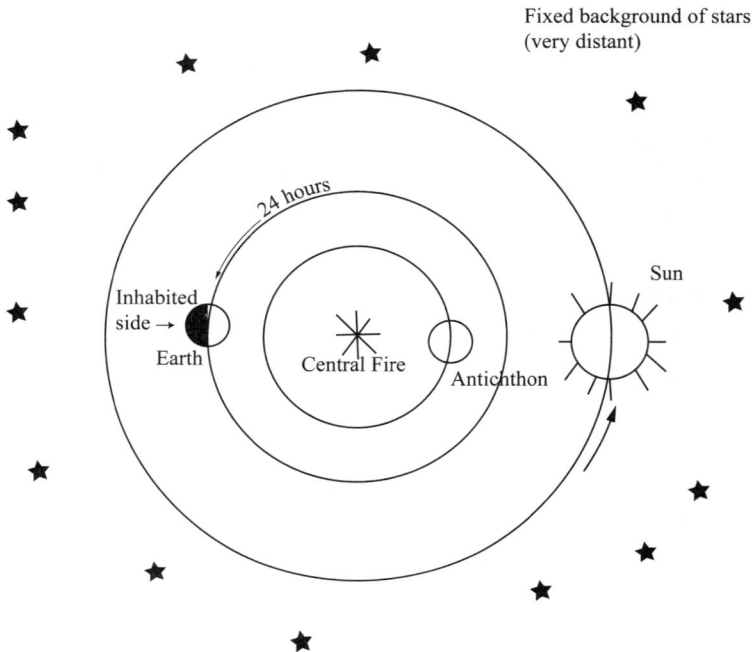

FIGURE 2.4 The universe of Philolaus. Both the Earth and Sun orbit a common center called the Central Fire. The inhabited half of the Earth always faces away from the Central Fire but orbits it daily, illuminated by the Sun during the daylight hours and turned away at night. Here, the inhabited half of the Earth is facing away from the Sun, and it is night.

perspective than that of Anaximander, who had the Earth motionless at the center of his universe.

While Philolaus's model explained how the stars move around the Earth on a twenty-four-hour cycle, the idea that this was caused by the Earth's rotation was also something that Pythagoreans believed. One, named Hicetas of Syracuse, is quoted as saying that the Earth, "while it turns and twists itself with the greatest velocity around its axis, produces all the same phenomena as if the heavens were moved and the Earth were standing still."[5] This concept doesn't restrict the Earth to be at the center of the universe. One can conceive of models like this with the Earth at the center or not.

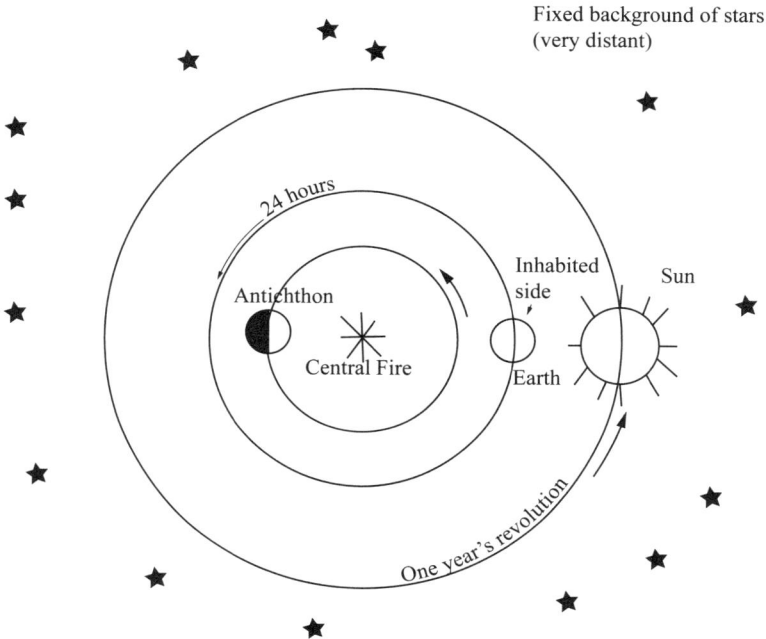

FIGURE 2.5 Twelve hours later, the inhabited side of Earth is facing the Sun and it is daytime there. The Sun orbits the Central Fire with a period of one year. This motion "explains" its eastward passage against the background of stars.

After the Pythagoreans, ancient Greek astronomy gained sophistication as more and more astronomical phenomena were catalogued. The depth of these is beyond the scope of this chapter, but the astronomers sought to account for a myriad of details. First, the motion of the planets garnered attention, but their motions seemed erratic. Like the Sun, the planets move roughly eastward against a fixed background of stars with a more-or-less steady motion, but at different speeds. The Sun moves to the east at approximately one zodiacal sign (30°) per month or one degree per day. The Moon, on the other hand, moves much more swiftly at about 12 degrees per day. Jupiter, on average, moves east at about one-tenth of a degree per day. Saturn is the slowest planet with a period of about twenty-nine years, and Mercury the fastest with a period of eighty-eight days. Unlike the Sun and

Moon, the planets will sometimes undergo *retrograde* motion where they appear to move west for some time and then regain their eastward path. Describing these and other phenomena resulted in increasing complexities of the models of space. In particular, the idea of the Earth at the center of space emerged, and the motions of the celestial objects were described as spheres. But spheres within those spheres were used to correct for the seemingly errant motions such as retrograde.

The fame of ancient Greek astronomy was realized in two key figures: Aristotle and Ptolemy. Aristotle's model was more notional and carried over into core beliefs of the Catholic Church. Ptolemy's model was more calculation driven and was the basis of almanacs of celestial motion produced well into the start of the Scientific Revolution.

SPACE, ACCORDING TO ARISTOTLE

Aristotle's (384–322 BC) model of the universe could best be described as egocentric. The Earth sits at the center, motionless, in much the same way Anaximander's universe had a flat cylinder at the center. While it might be an understatement to call Anaximander's vision of space "simple," Aristotle's was quite complicated. His model is illustrated in figure 2.6.

First, the Earth is a sphere, which is half occupied (the element Earth), and the other half an unoccupied ocean (the element Water). This is surrounded by the sphere of the element Air, which is then surrounded by the sphere of the element Fire.

If we use the "fixed" background of stars as a guide, there is an ordering of the speed with which the celestial objects move against this background. From slowest to fastest we have:

Saturn
Jupiter
Mars
Sun
Venus
Mercury
Moon

Prime Mover

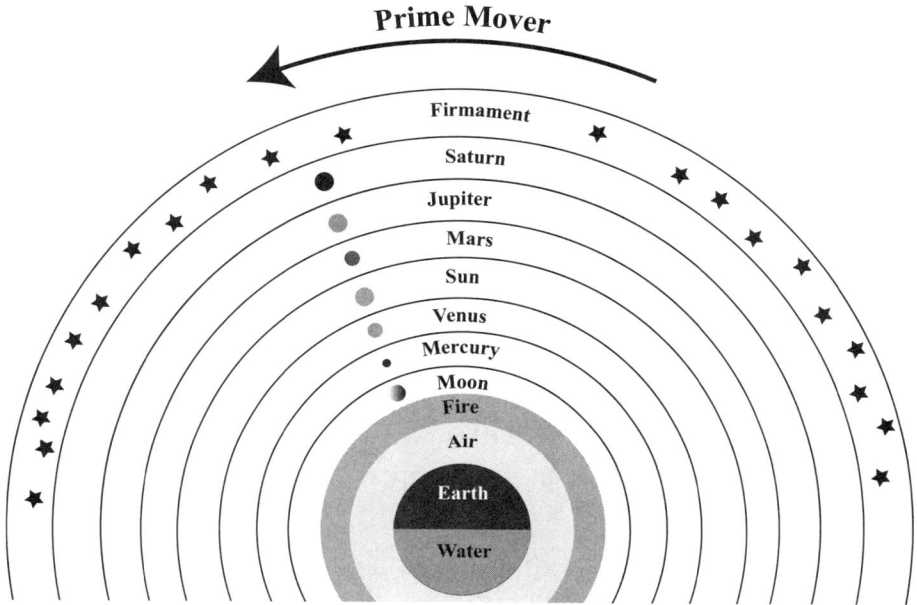

FIGURE 2.6 Aristotle's vision of space with a series of nested spheres representing the stars and seven celestial objects. The Prime Mover impelled the motion of the Firmament, which impelled the motion of the celestial objects, the force being transferred into each sphere until it reached a motionless Earth. Note that the figure does not capture the full complexities of the model.

The orbit of the Sun around the Earth corresponds to precisely a year. This is taken from the vantage point of our modern allocentric model of the solar system with the Sun at the center, and the background of stars acting as a frame of reference. Aristotle turns these motions upside down by insisting on an immobile Earth at the center of the universe. One argument made in favor of a stationary Earth is that a rotating Earth would produce wind speeds of 1,000 miles per hour at the equator. One hears echoes of this argumentation from flat-Earthers.

How do you answer the question of wind speeds on a rotating Earth? The easiest is to say that everything rotates together, whether it's land, the oceans, or the atmosphere. Going into a bit more depth, there's a property of fluids called *viscosity*, which is a measure of the "stickiness" of a fluid. Think about honey coating a spoon versus water coating a spoon. Since the honey is stickier than water, more of it clings to the spoon. Air is a fluid and has its own viscosity. It "sticks" to the surface of the Earth, which pulls it along. Since there is no surface in space that the air can stick to, it moves along with the surface of the Earth.

With an immobile Earth, the twenty-four-hour rotation of the firmament is the fastest. Saturn's rotation around a nonrotating Earth is just a little bit slower because it's moving eastward across the background of stars. Jupiter is a little slower still, then Mars, and so forth until we get to the Moon, which appears to move the slowest, and finally there is Earth.

How did Aristotle account for the motions in the sky? First, there is the concept of a "Prime Mover." This is an impelling force that causes the cosmos to rotate at a period of twenty-four hours. Just inside the Prime Mover is the Firmament, which is the sphere of the fixed stars. The Firmament transfers its motion to the next sphere, that of Saturn.

Motion is transferred down the chain from Saturn to Jupiter to Mars, and so forth. At each transfer, the speed of the successive spheres gets slower and slower until one reaches the stationary Earth. This seems like an odd way to look at the motions in the sky, but at some level the Moon, which would be slowest in this system, does move across the sky the most slowly when seen from the perspective of a motionless Earth.

The odd motions of retrograde and other phenomena were accomplished with spheres within the celestial spheres, resulting in a total of fifty-five spheres to account for all the observed motions.

There are interesting hints of modern science in some of the ancient Greek philosophers but these were overshadowed by Aristotle's outsized influence. Atomist philosopher Democritus held that the Milky Way consisted of stars so distant that they blurred into a continuum. Aristotle, on the other hand, believed that both the Milky Way and comets were a kind of atmospheric phenomenon that were the result of contact between the celestial spheres and the outermost terrestrial sphere of fire. Perhaps it doesn't need to be said, but Democritus was correct and Aristotle was wrong.

It's worth pausing here to examine the influence of Aristotle's universe. Of all the models, this one influenced Western culture for 2,000 years. Why? Part of this had to do with the persistence of his writings. More importantly, perhaps, is the wedding of Plato's ideas to Aristotle's model. Plato asserted that there were perfect forms that were eternal and true, and human existence is just a corrupt, malleable manifestation of the true and eternal forms. The Catholic Church adapted the worldviews of Plato and Aristotle, and this is perhaps the most prominent reason for the persistence of his model.

With the Firmament and Prime Mover added, that leaves nine perfect spheres of heaven. This idea was developed into the concept of a hierarchy of angels that dwelled in the spheres of heaven. An early Christian writer known as Pseudo-Dionysus wrote a number of treatises, including *The Celestial Hierarchy* and the *Ecclesiastical Hierarchy*. In his writing, he posited nine kinds of angels, with the most perfect, the Seraphim, occupying the sphere of the Prime Mover, and then the Cherubim (angels with wings) occupying the firmament, and so forth until we have the lower Archangels occupying the sphere of Mercury, and the Lesser Angels in the sphere of the Moon. Thomas Aquinas echoed this hierarchy in his work *Summa Theologica*, including the division between Earth and the heavenly realms. At some level, this aligns well with the notion that a hierarchy of beings exists as its own map. Here, a hierarchy of God and angels exists in this version of the universe. The centrality of Aristotle's universe to the Catholic Church cannot be overstated and, seen from this perspective, it is understandable why they would be reluctant to part with it for the Copernican model with the Sun at the center.

Although one might naively think that Aristotle gave a kind of primacy to the Earth by having it at the center of the universe, that he elevated humans in some way, the opposite was the case. Earthly things were considered in the realm of sin and imperfection, while the heavens were more perfect. The Prime Mover may be thought of as representing God, with the inner spheres reflecting less and less perfection.

This ordering is reflected in the word *sublunary*. A quick check of the definition of "sublunary" yields: "belonging to this world as contrasted with a better or more spiritual one." More to the point, in Aristotle's model, the spheres of the Moon and beyond, the heavens, were perfect and eternal, while the corrupt and malleable sphere of Earth lay inside the sphere of the Moon. In Christianity the idea of an imperfect body and an eternal soul match well with the dualism of an imperfect sublunary Earth and a perfect heavenly realm.

The question of astrology also intrudes here. If the Prime Mover imparts all motion, including motion on Earth, the idea that the stars and planets transmit influence to the world is not a big stretch of the imagination. On the other hand, the Catholic Church views astrology as a kind of desire for power and knowledge that rightfully belongs to God. Nonetheless, the idea of a transmission of motion and influence from the heavens to Earth persists in astrology, which was quite popular through the Middle Ages and well beyond.

What exists outside the Prime Mover? In the Aristotelian model there is no space and time beyond.

THE UNIVERSE OF ARISTARCHUS

Aristarchus of Samos (310–230 BC) put forth a model of space that was allocentric and was remarkably modern when seen in retrospect. He placed the Sun at the center, has the Earth orbiting the Sun, and the Earth rotating. Like the model of Philolaus, this has the advantage that the distant stars aren't moving at a huge speed, rotating once every twenty-four hours. There is also a straightforward explanation for the passage of the Sun eastward against the background of stars.

The Greek mathematician Archimedes (287–212 BC) wrote a treatise called *The Sand Reckoner*, about his calculations of how many grains of sand it takes to fill the universe. In it he used Aristarchus's model as a way of estimating the size relative to the size of the Earth. Here is a translation of a passage from *The Sand Reckoner* that refers to Aristarchus.

> Now you are aware that "universe" is the name given by most astronomers to the sphere whose center is the center of the Earth and whose radius is equal to the straight line between the center of the Sun and the center of the Earth. This is the common account, as you have heard from astronomers. But Aristarchus of Samos brought out a book consisting of some hypotheses in which the premises lead to the result that the universe is many times greater than that now so called. His hypotheses are that the fixed stars and the Sun remain unmoved, that the Earth revolves about the Sun in the circumference of a circle, the Sun lying in the middle of the orbit, and that the sphere of the fixed stars, situated about the same center as the Sun, is so great that the circle in which he supposes the Earth to revolve bears such a proportion to the distance of the fixed stars as the center of a sphere bears to its surface.[6]

Archimedes first refers to "most astronomers" being those who adhere to the Aristotelian model, then pivots to the model by Aristarchus.

The one surviving writing of Aristarchus concerns the relative size of the Sun and Moon and their respective distances from Earth. From his measurements and reasoning, Aristarchus believed that the Sun was nineteen times larger than the Moon, and substantially larger than the Earth. Although this is a huge underestimate in modern terms, Aristarchus took the size of the Sun to be so large that it must be the center of the universe.

We don't know whether Aristarchus had a more complete model of the solar system with the other planets orbiting the Sun, but it's certainly quite possible.

PTOLEMY

Both Aristotle's universe and Ptolemy's universe persisted through the Middle Ages into the seventeenth century. While the audience for Aristotle was more for the clergy, the audience for Ptolemy was more for astrologers. Aristotle's model was notional, Ptolemy's model was calculational.

Ptolemy (100–170 AD) lived in Alexandria, Egypt, and was, in a sense, the "last word" on ancient Greek astronomical knowledge. He wrote famous treatises on geography, astrology, and even music. His most famous work in the modern era is the *Almagest*, or *Almanac*, which contained his prescriptions for how to calculate the positions of celestial objects in the sky, along with tables that were the result of his calculations.

Astrology was a major driver of the creation of astronomical tables dating back at least to the Babylonians. It takes astronomical data to construct astrological charts. This, in turn, requires an almanac that gives the positions of the celestial objects. Making the almanac requires a schema of calculations, and this was what Ptolemy developed. In his model, Ptolemy had the same ordering of the surrounding spheres of the Sun, Moon, and planets as in Aristotle's model, but the Earth wasn't precisely at the center.

Figure 2.7 illustrates Ptolemy's model. There is a large orbital circle called a *deferent*. The center of a circle, called the *epicycle*, orbits about the center of the deferent. The celestial object (Sun, Moon, or one of the planets) have motions described by it rotating around the center of the epicycle while the center of the epicycle itself orbited around the center of the deferent. To make everything work out properly, the Earth had to be displaced by some amount from the center of the deferent, called the *equant*. So, technically speaking, the Earth wasn't the center of the universe for Ptolemy's model, although it would be fair to call Ptolemy's universe geocentric.

The Sun was something of a special case. Compared to a perfectly circular motion around the Earth, the Sun appears to speed up and

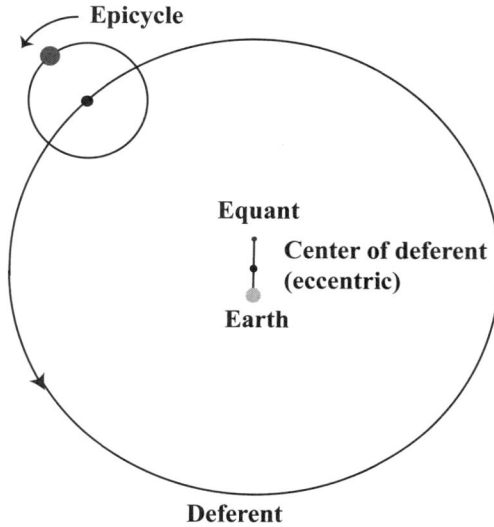

FIGURE 2.7 Ptolemy's construction for calculating the positions of planets and the Sun and Moon.

slow down over the course of a year. In modern terms this is partly the result of eccentricity of the Earth's orbit. We view the planetary orbits as ellipses, which are effectively distorted circles, where the amount of distortion is called its *eccentricity*. To get this motion in his model for the Sun Ptolemy used either the epicycle or the equant but did not need both.

On the other hand, the motion of planets is more complicated. In modern terms, when the Earth overtakes a *superior planet* (Mars, Jupiter, Saturn) or an *inferior planet* (Mercury, Venus) overtakes the Earth, the motions of the planets appear to go backward for a period: *retrograde motion*. Both the equant and epicycle were needed to account for the additional wrinkle of retrograde motion. I would label Ptolemy's system egocentric, even though the Earth is not at the precise center of the deferents. It has a fixed Earth and the sky is turning around it.

It turns out that any closed curve can be traced out with a suitably large number of epicycles and equants, so Ptolemy hit on a modern mathematical model that allowed him to fit the orbits of the celestial objects as seen from Earth.

Almagest was dutifully transcribed multiple times, first to Arabic, and then into Latin in the twelfth century. Many Arabic scholars in the Middle Ages built upon Ptolemy's work, including updating his treatise on geography.

If you feel Ptolemy's model seems complicated, you aren't alone. In many ways, it's more of a fit to the planetary motions than it is a physical model, like Aristotle's. By this I mean that Aristotle's model had forces transmitted from a Prime Mover on down through each of the celestial spheres. On the other hand, Ptolemy had mostly a geometric construction that enabled calculations. Nonetheless it was precise. Ptolemy's method of calculation persisted until Johannes Kepler in the seventeenth century. The heliocentric model we associated with Copernicus seems much simpler than something with epicycles and eccentrics, but it took until Kepler to demonstrate a superior set of calculations for the almanacs with a Sun-centered system. One of the challenges in a heliocentric model is translating the motions into what we see from Earth. Although Aristarchus seems to have gotten the elements correct, the translation from the allocentric perspective to what we see on Earth is not so easy, but there are clues along the way as to why the heliocentric model makes more sense. Next, I tackle some of the important elements in this translation.

* 3 *

IMAGINING *and* FINDING
the PLANETS

As we saw in the previous chapter, the model of Aristarchus, with the Sun at the center of Earth's orbit, is remarkably modern in contrast to the more egocentric models of Aristotle and Ptolemy. Making the transition to our heliocentric system requires visualizing the motions of planets in what might seem like a simple model with the Sun at the center and circular orbits of the planets. The motion of the Earth, however, complicates the shift into an egocentric perspective.

Recently, a Facebook friend who is also a physicist posted a question there, "What is that bright star that I see in the western sky just after sunset?" It was Venus. It's rare for people to identify bright stars and constellations, and the wandering planets are more challenging still. But to appreciate the shift in perspective, the exercise of imagining the behavior of the planets and then going the extra mile to identify and track them in the sky is enlightening. It puts you in the same position as the early astronomers. Here I will focus on three main features to identify planets and their motions.

Learning about the solar system is standard curriculum in grammar school, and many of us have made crayon drawings of the Sun and planets that look something like figure 3.1. We trace out the classic

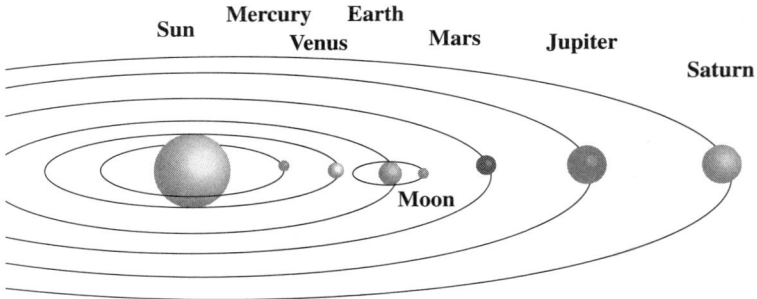

FIGURE 3.1 The modern version of the solar system as taught in grammar school. I omit the planets beyond Saturn, as these aren't visible to the naked eye.

allocentric model of the solar system with the Sun at the center and the familiar pattern of Mercury, Venus, Earth, Mars, Jupiter, and Saturn orbiting in nearly circular paths.

In this exercise of the imagination and planet-gazing I focus on three features: (1) planet brightness or *apparent visual magnitude*, (2) planet angle from the Sun, or *elongation*, and (3) planetary position against the stars or motion along the *ecliptic*. These features were available to ancient Greek astronomers and are also observable characteristics that the modern stargazer can key into. If you do get out and search for the planets in the sky, observing these characteristics requires some patience and the willingness to track planets over the course of at least a year.

APPARENT VISUAL MAGNITUDE

How bright a star or planet appears to our eyes is known as "apparent visual magnitude." It's graded on a scale where less is more. The larger the number, the dimmer the star. The numbers even run into the negatives. For example, Sirius is the brightest star in the sky and has an apparent visual magnitude of −1.5. I won't go into too much detail about stargazing here, but Sirius can be seen just to the east of the constellation Orion, which is high in the sky after sunset during the winter months.

In contrast to Sirius, Polaris is substantially dimmer but has the role of the pole star. It has an apparent visual magnitude of 2.0. It's dim enough that it can sometimes be a challenge to see in areas with light pollution. Splitting the difference between Polaris and Sirius is the star Vega. It's found in the constellation Lyra and can be seen after sunset on summer nights. Vega has an apparent visual magnitude of 0. So, between Sirius (−1.5), Vega (0), and Polaris (2) we have a rough calibration of the apparent visual magnitudes for our eyes.

In *Almagest* Ptolemy created a star-catalog of approximately a thousand stars, giving their positions in the sky and apparent visual magnitudes based on his system. Our modern system represents an attempt to roughly mimic Ptolemy's classification. For the purposes of this discussion, I'll take "brightness" to be synonymous with "apparent visual magnitude." The discussion here omits some details.[1]

Perhaps it's obvious, but the brightness of the planets is due to reflected light from the Sun. What we see as the apparent visual magnitude is a product of the distance to the planet, its size, and even phases.

Saturn is the dimmest observable planet with an apparent visual magnitude of roughly 0, making it on par with the brightness of Vega. Because it's so far away from us in its orbit, our relative distance to Saturn doesn't change much over the course of a year. Accordingly, its brightness is roughly constant over time.

Jupiter is the brightest of the superior planets (outside Earth's orbit), with an apparent visual magnitude that varies from −3 to −1.7, with an average of about −2.2. This makes it brighter than Sirius even at its dimmest. The variation in brightness is larger than Saturn as it's closer to us and the effect of our two orbits creates a more substantial change in relative distance. Part of the relatively high apparent visual magnitude is due to its size as the closest of the gas giants in the solar system.

We have the largest relative variation in our distance to Mars and, not surprisingly, this creates a large variation in apparent visual brightness. This varies from -2, when we're closest to Mars in our respective orbits, to 2 when we're on opposite sides of the Sun. With this variation, it can be as bright as Jupiter, or dimmer than Saturn, depending on where we are in our respective orbits. One time, when at our point of closest approach to Mars, it was so bright that I could see its reflection on the waters of a calm lake in northern Minnesota.

You might think that the variations in brightness of a planet like Mars might be a clue that we have separate orbits around the Sun. But, with the epicycles in Ptolemy's model, you can still imagine the distance shifts that could account of a variation in brightness.

The brightness of Venus is where things get interesting. For the superior planets, the light mostly reflects directly off the surface or atmosphere, and there's not much to talk about except the sizes of the planets and their distances from Earth. The inferior planets of Mercury and Venus go through a full set of phases, like the Moon. When Venus is opposite the Earth in its orbit—that is to say, the Earth is on one side of the Sun and Venus is on the opposite side—the full face of Venus is illuminated. On the other hand, when Venus overtakes us in our orbit, it presents more of a dark side to us. The combined effect of the phases and changes in distance conspire to make the apparent visual magnitude nearly constant. Venus is the brightest celestial object after the Sun and Moon and has an apparent visual magnitude of -4.

Without knowing about phases and the changing distance to other planets, it's difficult, however, to reconcile the variation in Mars's brightness with the rough constancy for Venus. The issue of the changing brightness of the planets was indeed noticed in Ptolemy's era, and his model with epicycles and eccentrics at some level explained it, but the difference between the shifting brightness of Mars and the constancy of Venus did not really get solved.

The phases of Venus were first observed by Galileo through a telescope in the seventeenth century. This observation by itself lent considerable weight to the Copernican model of the solar system.

Finally, we have Mercury, which has a huge range in apparent visual magnitude. This is one of a few reasons why it's challenging to spot. On occasion it can be almost as bright as Jupiter, or much dimmer than Saturn. Its changes in phase are one cause for such a strong brightness variation.

ELONGATION

The angle between the Sun and a planet in the sky as seen from the Earth is known as "elongation." The point of maximum elongation is when this angle is the largest. The superior planets of Mars, Jupiter, and Saturn can wander to any angular distance from the Sun, so we could see elongations as large as +180° or −180°. When Mars is at 180° from the Sun, we are at the point of closest approach in our orbit, and Mars is at its brightest with a magnitude rivaling Jupiter's.

On the other hand, the inferior planets of Mercury and Venus cannot stray any farther than a certain elongation on account of their orbits, as can be seen in figure 3.2. The points of greatest elongation give the best chance of seeing the planet. The maximum elongation of Venus varies from 45° to 47°. Mercury's maximum elongation varies between 18° and 28°. The range of maximum elongations in both cases arises because the planetary orbits are ellipses, not perfect circles.

Because Venus stays close to the Sun, it's often seen as the morning star when it rises before dawn and the evening star when it sets after sunset. So if you see a very bright star in the sky near where the sun rises or sets, the odds are excellent that it's Venus.

Mercury's elongation and its apparent visual magnitude conspire to make it difficult to see. With small elongations, it's often too close to the Sun in the sky to be visible, due to the Sun's brightness. The best time to spot Mercury is at its point of maximum elongation. Even then, it takes a bit of care to find it. I can count on one hand the number of times I've seen Mercury.

Returning to ancient planetary models, the elongations of Mercury and Venus, that is, their tendency to be close to the Sun in the

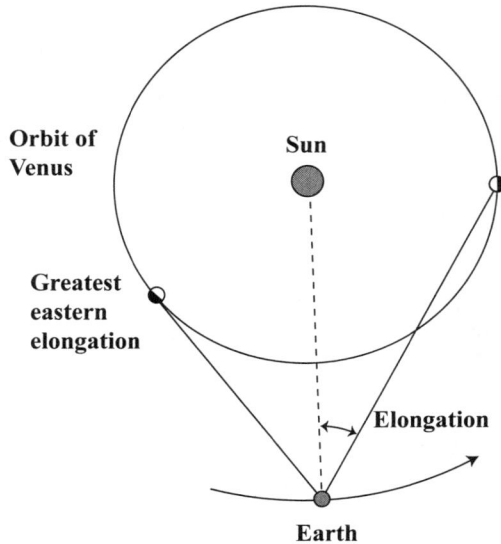

FIGURE 3.2 Elongation of a planet is the angle in the sky from the Sun to the planet. Shown here is the elongation of Venus when it is at a point when it's perpendicular to a line drawn from the Earth to the Sun. Also shown is the point of its greatest elongation in the east—just before sunrise. Finally, I've put in the phases of Venus to show the illuminated side facing the Sun.

sky, could have been yet another clue that the Sun was the center of the solar system but was missed. Clearly Ptolemy had a system that "worked" in the sense that it predicted the positions of Mercury and Venus with some accuracy, but once we have a Sun-centered system, things seem so much simpler and perhaps obvious.

POSITION IN THE ECLIPTIC

Two things are happening in Earth's motion about the Sun. First, we orbit the Sun with a period of one year. From Earth's perspective the Sun moves east against the background of stars. Another motion arises from the tilt of Earth's axis. The Northern Hemisphere is tilted toward the Sun in the summer and away in the winter. As a result the Sun is high in the sky in the boreal summer and low in the

Earth's rotation axis

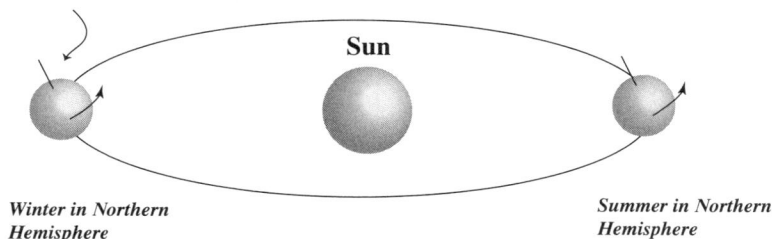

FIGURE 3.3 From a heliocentric perspective, the Earth's seasons come from the combination of the axial tilt and its orbit about the Sun.

boreal winter, for us inhabitants of the Northern Hemisphere. The effective latitude in the sky of the Sun is called its *declination*. This is illustrated from the allocentric perspective of our orbit about the Sun (see fig. 3.3).

Shifting to our perspective on Earth, we can understand how the effect of axial tilt and our orbit conspire to produce the path of the Sun in the sky. Figure 3.4 illustrates the concept of a celestial sphere and how the Sun moves through the sky over the course of a year.

The celestial sphere is a way of projecting the positions of objects onto the sky as seen from Earth. Imagine a sphere concentric with our Earth but floating out in space above us. Directly above and below the North and South Poles of Earth are the celestial North and South Poles. Projecting out from the equator is the celestial equator. The angle above or below the celestial equator is called the *declination* of a celestial object. While declination is the analog of latitude on Earth, astronomers use a slightly different convention for the analog of longitude. This is not worth getting into here, but the effective celestial longitude is based on the fixed background of stars for reference. The "prime meridian" of the celestial longitude is taken, by convention, to be the point where the Sun crosses from the southern to the northern celestial hemisphere at the vernal equinox. The path of the Sun is called the *ecliptic* and is shown on the celestial sphere in figure 3.4.

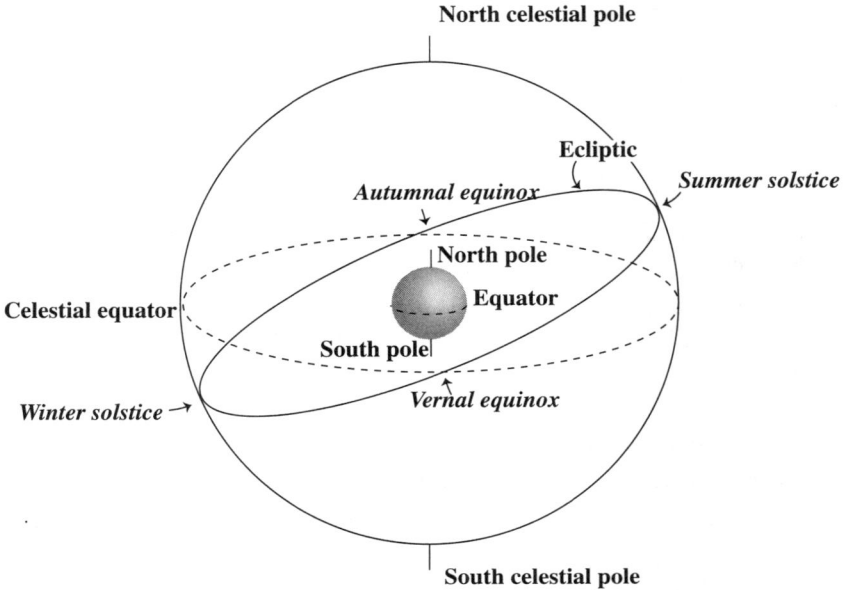

FIGURE 3.4 The solar ecliptic as seen from a geocentric perspective.
Over the course of the year the Sun will move eastward against
the background of stars and will change declination.

The Sun has a declination of zero at the vernal equinox and autumnal equinox. It gains the highest declination at the summer solstice and the lowest declination at the winter solstice.

Now, what does this have to do with planets? Owing to a lucky break of nature, the solar system has planetary orbits that are contained in a plane. Since the Earth's orbit is in that plane, the ecliptic also represents the solar system projected onto the celestial sphere. The planets are thus confined to move along the ecliptic, just like the Sun, but with different speeds.

How does this help you locate the planets? For starters, you need to be able to work your way through the ecliptic by viewing the stars. Locating Venus is straightforward. It will always be relatively close to the Sun, and since the Sun is in the ecliptic, it will appear in the west after sunset and in the east before sunrise.

For Mars, Jupiter, and Saturn, it's a bit more guesswork as these planets wander all over the ecliptic. My advice is to start with either a written guide to the stars and constellations or to get an app and learn how to locate them.[2]

Here are some major stars and constellations that lie in or near the ecliptic that can be easily identified: The Pleiades (Seven Sisters) are a fuzzy patch of stars. Just to the east of the Pleiades is the constellation Taurus the Bull with the distinctive yellow-orange star Aldebaran. Further to the east is Gemini with the twin stars Castor and Pollux. To the east of Gemini is the constellation Leo with the blueish star Regulus. East of Regulus is Virgo, which contains the bright star Spica. Finally, Scorpio is to the east of Spica with the orange-colored star Antares in the middle of the body of the scorpion. Other constellations along the ecliptic can be more difficult to identify.

RETROGRADE MOTION

One of the challenges in Ptolemy's construction was getting retrograde motion correct with the epicycles and eccentrics. In our modern heliocentric version of the solar system, the seeming backward motion of planets is explained by the Earth overtaking a superior planet, like Mars, or an inferior planet overtaking us in our orbit, like Venus.

Figure 3.5 shows retrograde motion of Mars from both the allocentric perspective of the Earth overtaking it in its orbit, and the egocentric perspective of what we see on Earth. Mars moves mostly west to east until it enters retrograde and then appears to travel east to west until it eventually regains the west-to-east motion.

Mars is the best planet for observing retrograde motion because of the geometry of our respective orbits. Since we overtake Mars when we're near the point of closest approach, it will be quite bright and visible in the night sky opposite the Sun. Retrograde of Mercury and Venus are difficult because they're so close to the Sun when they overtake us in their orbits.

Background of fixed stars

A   B   C

Mars orbit

B

C   **A**

Earth orbit

B

East   **A** ────→ ────→ **C**   **West**

Apparent motion of Mars

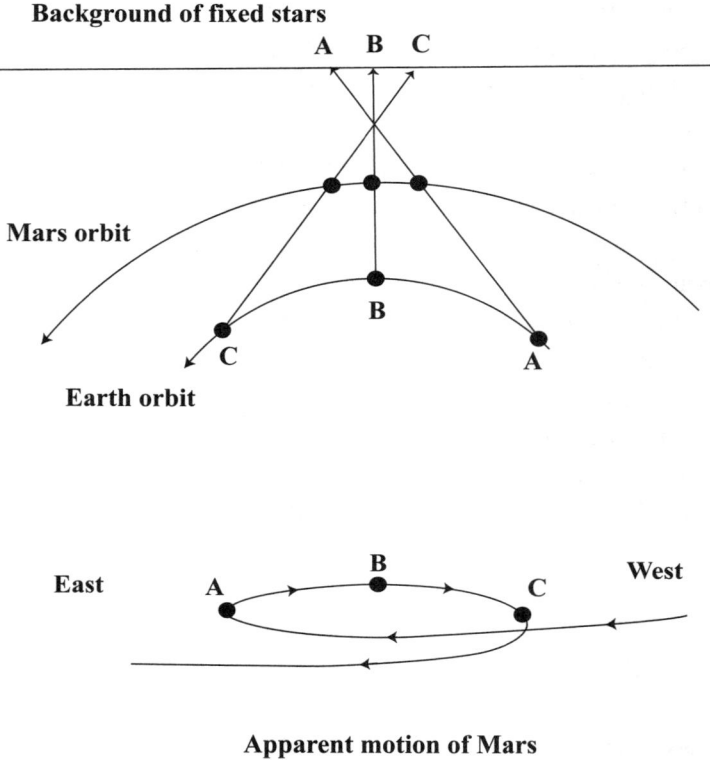

FIGURE 3.5 Retrograde motion of Mars as seen from an allocentric perspective (*top*) and from an egocentric perspective (*bottom*).

PRACTICALITIES

One pointer in viewing the planets is the steadiness of their light. Even though planets seem like stars to the naked eye, they don't twinkle like stars. Part of this has to do with the fact that stars are effectively point-like objects, but the planets have a finite diameter. Dust and turbulence in the atmosphere can cause stars to twinkle and shift color, while the light from the planets is steadier.

Two planets are relatively easy to spot: Venus and Jupiter. Both are the brightest objects in the sky after the Sun and Moon. Repeating myself: Venus will be close to the Sun, either appearing before sunrise or after sunset. Jupiter will wander, so (except for the relatively rare cases when it's close to the Sun) the brightest object in the sky away from the Sun is almost certainly Jupiter.

Mercury is difficult because it's both faint and close to the Sun. You will be lucky to catch it.

Saturn and Mars can present more of a challenge because they can be faint and look like an ordinary star at times. When we are close to Mars its red coloration and brightness makes it distinctive. If your stargazing skills allow you to trace out the ecliptic, this improves your chances of catching Saturn or Mars.

Once when I was in Jamaica on vacation, I was stargazing and looked directly overhead to see the constellation Taurus. I was in the process of picking out the orange-colored star Aldebaran, the eye of the bull, when I saw what looked like *two* versions of Aldebaran, both red-orange and roughly the same magnitude. It took me a minute until I realized, "Hey, Taurus is in the ecliptic, and that other star is reddish. . . . It must be Mars!"

Another time I was looking south toward Scorpio from my house on Cape Cod. I found the orange-colored star Antares, but something didn't seem right. There was another star in the familiar formation. Again it took me a minute, again I realized, "Oh yes, Scorpio is in the ecliptic, it's not red like Mars, it's not so bright, it must be Saturn!"

I had one lucky stargazing moment when all the visible planets, the Sun, and the Moon were closely aligned in space. This was on the morning of June 16, 2022, in the predawn hours looking east. The planets were aligned in the order of Mercury, Venus, Mars, Jupiter, Saturn, and the Moon (fig. 3.6). The Sun was a glow on the horizon It was quite remarkable and was just an accident as one of my dogs woke me to let her out to do her business in the wee hours.

The speed of the planets across the sky and their other characteristics made it inviting to project human-like characteristics on to them. This habit appears in language. We speak of a *saturnine* disposition of one who is slow and gloomy. A *jovial* person is upbeat and happy. *Martial* law can be declared in an emergency. We might catch

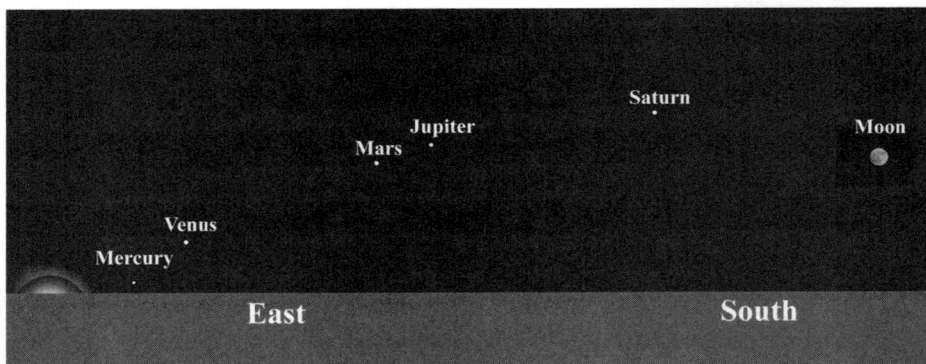

FIGURE 3.6 Alignment of the Sun, planets, and Moon on
June 16, 2022. All of them are in the ecliptic.

a *venereal* disease from sex. If you change your mind a lot, you're *mercurial*. Maybe you have a *sunny* disposition, or you're a raving *lunatic*.

When I saw Mars in Taurus, it made me think of war, seeing it next to the orange-red Aldebaran. While my magical thinking on this observation was short-lived, many people order their lives through astrology. I will often say that "Mercury must be in retrograde" when I have difficulty communicating. This habit of projecting human drama into the sky goes back at least to the Babylonians, but almost certainly it goes back much further in time.

* 4 *

MERCURY MUST BE
in RETROGRADE

Astrology is one of the most enduring and global means of divining and is built on an explicit linkage between the heavens and human affairs on Earth. If we look back to Mesopotamia, you could say that Western astronomy got its start with astrology. Ancient Greek astronomy added additional layers with the knowledge of the motions of the planets. But I should be clear: astrology is a cultural phenomenon based on models of space.

There are multiple variations of astrology. For example, there is a Hindu branch of astrology that is related to the studies of sacred Vedic texts. Here I'll focus on the practice of Western astrology that grew out of the Greek tradition, specifically Ptolemy's version. We saw in chapter 2 the advent of Ptolemy's system for calculating planetary positions, in his work *Almagest* (Almanac). In his work *Tetrabiblos* (Four Books), Ptolemy lays out his rules for astrology, which involve spatial constructions that are partly built from his geocentric model.

In the beginning of *Tetrabiblos* Ptolemy lays out his case for the efficacy of astrology in predicting events. As a scientist I should scoff at this notion, but there *is* some causality associated with events in the sky to those on Earth. Ptolemy uses this as a jumping-off point. The influence of the Sun is undeniable. It gives us days, the year, and seasons. It's also considered something of a ruler or king of the heavens.

The influence of the Moon is a bit more subtle, but it influences events on Earth. It's responsible for the tides in the oceans, including biological processes, like the behavior of oysters.[1] We sometime make quips about "it must be a full Moon," when someone acts a bit off. Sleep habits are sometimes disrupted. Then, there are events associated with the heavens that aren't causations but *are* correlations. In chapter 2 I described how the heliacal rising of Sirius foretold the flooding of the Nile. Likewise, Hesiod wrote that the heliacal rising of the Pleiades heralded harvest time in ancient Greece. Arguing from undeniable influences on Earth, like the Sun and Moon, Ptolemy then ascribes the heavenly influences to the planets and beyond.

SIGNS OF THE ZODIAC

What is the space of astrology? We can start with the Zodiac. The signs of the Zodiac are a way of subdividing the ecliptic into twelve angular swaths of 30 degrees each. Figure 4.1 shows this subdivision,

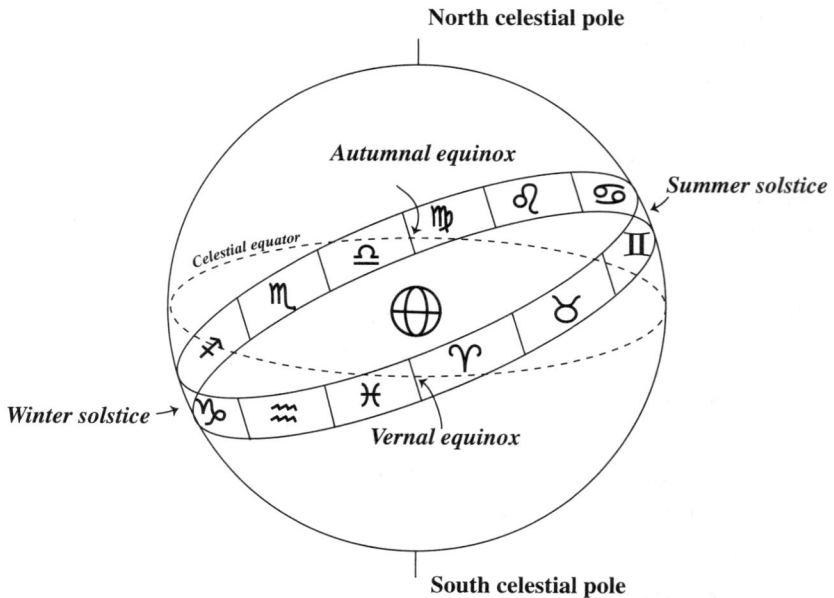

FIGURE 4.1 The signs of the Zodiac ringing the ecliptic.

and if you look back at figure 3.4 in the previous chapter, you can see how the two are basically the same concept of celestial sphere projecting out from the Earth. Why twelve angular swaths? The number 12 has 2, 3, 4, and 6 as common divisors, and there is a system to assign each Zodiac sign with a unique character.

Like figure 3.4, the beginning of the Zodiac is at the point where the Sun crosses from the Southern to Northern Hemisphere at the vernal equinox. This is the boundary between Aries, the first sign, and Pisces, the last. The signs then go in order as shown in figure 4.2.

The Zodiac signs are associated with the names of constellations found in the ecliptic. One perhaps minor point is that the physical constellations no longer align with the Zodiac signs. For example, you might read on an astrology website that Jupiter is in Taurus, but when you look in the sky, you see it in Aries. The assignment of Zodiac signs dates from the time of Ptolemy when the signs lined up with the constellations. The Earth's rotational axis wobbles with a period of about 26,000 years, meaning that the position of stars in the sky relative to the ecliptic has shifted since the time of Ptolemy. Nonetheless, astrology uses the Zodiac signs from roughly 200 AD. This is a known feature.

♈ Aries
♉ Taurus
♊ Gemini
♋ Cancer
♌ Leo
♍ Virgo
♎ Libra
♏ Scorpio
♐ Sagittarius
♑ Capricorn
♒ Aquarius
♓ Pisces

FIGURE 4.2 Symbols associated with the signs of the Zodiac.
These can be seen in figure 4.1 along the ecliptic. The ordering
around the ecliptic goes from top to bottom.

In Ptolemy's schema for calculating the positions of the Sun, the Moon, and the Planets, the locations are indexed to how many degrees they are into each Zodiac sign. So, we might say that Jupiter is 12 degrees into Taurus.

SUN SIGNS AND CHARACTER OF THE ZODIAC SIGNS

If you open a newspaper, or look online, you'll see the horoscope giving predictions for a given "sign." In the 1960s and early '70s, astrology experienced a revival and there was a standard pickup line of "What's your sign?" This is referencing the position of the Sun in the Zodiac at the time of your birth. So, if you were born on August 1 the Sun was in Leo, and you'd be "a Leo." People who know you might utter, "Such a Leo." This brings us to the method for assigning the character to Zodiac signs.

Masculine/Feminine

There is a two-fold alternation across the Zodiac, starting with Aries as masculine, Taurus as feminine, Gemini as masculine, and so on, until each sign gets a masculine or feminine designation. In terms of what this means, the typical understanding is that masculine represents a kind of extroverted energy, projecting out into the world, with physical actions. The feminine characteristic is introverted and an energy that lies within, for example projecting inner emotional strength.

In this alternating schema, there are the masculine signs of Aries, Gemini, Leo, Libra, Sagittarius, and Aquarius. The feminine signs are Taurus, Cancer, Virgo, Scorpio, Capricorn, and Pisces.

Modalities

This is a three-fold division across the Zodiac signs and is related to the seasons. Each season has three Zodiac signs. For example, spring has Aries, Taurus, and Gemini, spanning the period from when the Sun is between the vernal equinox until Summer Solstice. So, each season has the following:

Spring: Aries, Taurus, Gemini
Summer: Cancer, Leo, Virgo
Autumn: Libra, Scorpio, Sagittarius
Winter: Capricorn, Aquarius, Pisces

The modalities of each sign depend on where they lie within each season. The start of the season is associated with the *cardinal* modality, which is a kind of initiating energy, like the start of a new project. The middle sign within each season has a *fixed* modality, which generally implies a stick-to-it kind of attitude, liking continuity and disliking change. The third sign in a season is the *mutable* modality, where there is an imminent change. For example, Pisces is late February, early March, and there are strong storms associated with the vernal equinox. Mutable modalities are flexible and can sometimes lose focus, in distinction to the fixed modality.

TABLE 4.1 Modalities and the associated Zodiac signs

Modality	Character	Signs
Cardinal	Active	Aries, Cancer, Libra, Capricorn
Fixed	Inflexible	Taurus, Leo, Scorpio, Aquarius
Mutable	Adaptable	Gemini, Virgo, Sagittarius, Pisces

Elements

Again, this is the Western tradition that borrows from the ancient Greek concepts, so the four elements of Fire, Earth, Air, and Water are involved.

The characteristics of each element are:

Fire: Action, creativity, transformative
Earth: Substance, practicality
Air: Intellect, the mind
Water: Emotions, receptivity, fluidity

TABLE 4.2 The elements and the associated Zodiac signs

Element	Character	Signs
Fire	Action, transformation	Aries, Leo, Sagittarius
Earth	Substance, practicality	Taurus, Virgo, Capricorn
Air	Intellect	Gemini, Libra, Aquarius
Water	Emotions, receptivity	Cancer, Scorpio, Pisces

The rotation in the Zodiac is to start with Fire for Aries, then Earth for Taurus, Air for Gemini, and Water for Cancer. It then repeats in this order for the other Zodiac signs.

So, putting it all together, we might have for my Sun sign of Pisces the qualities of female, mutable, and water. This implies that Pisces is introverted, adaptable, emotional, and receptive. You can read off your favorite sign from the tables and draw your own conclusions.

RULING PLANETS

At the time of Ptolemy there were seven celestial objects in the ecliptic. Each was associated with a god, giving the planets their characteristics. The common influences are shown in table 4.3.

There is a kind of "home" or chart ruler for each of the Zodiac signs. Unlike the gender, modality, or element, the association of planets to the Zodiac signs doesn't have an obvious rotation, since 12 is not divisible by 7. The general pairing in Ptolemy's schema is:

Aries–Mars
Taurus–Venus
Gemini–Mercury
Cancer–Moon
Leo–Sun
Virgo–Mercury
Libra–Venus
Scorpio–Mars
Sagittarius–Jupiter
Capricorn–Saturn
Aquarius–Saturn
Pisces–Jupiter

For the characters of the celestial objects, we can focus on Mars. It's traditionally associated with war and aggression. The red color of Mars resembles blood. This is an interesting coincidence, as the red color of blood comes from iron in hemoglobin, and the surface of Mars has much iron oxide that gives its coloration. The association of Aries with Mars has some logic to it. Looking at the characteristics of the signs, Aries, the sign of the Ram, is masculine, cardinal with the element of Fire, meaning active.

TABLE 4.3 Planets, their symbols, associated Greek gods, and influences

Planet	Symbol	Greek god	Influence
Saturn	♄	Cronus	Agriculture and harvest
Jupiter	♃	Jupiter/Jove	Growth, abundance, wisdom
Mars	♂	Ares	Strength, aggression, anger
Sun	☉	Helios and Apollo	Ego, purpose, vitality
Venus	♀	Aphrodite	Romance, lust
Mercury	☿	Hermes	Communications, wit
Moon	☾	Artemis	Nurturing, emotional life

Note: The ordering reflects the speed across the ecliptic from slowest (Saturn) to fastest (Moon).

In terms of motion in the sky, if a planet is moving on its normal west-to-east path across the ecliptic, events are associated with the positive influence of the planet. But, when a planet goes into retrograde with an east-to-west motion, it usually implies the opposite of the influence. So, if someone tells you that "Mercury is in retrograde," it means something like this: Mercury was the messenger of the gods and is associated with an influence of communication. If Mercury is in retrograde, the implication is that communication breaks down or is difficult.

THE ASTROLOGICAL WEEK

Another cultural manifestation associated with astrology is an explanation for the ordering of the seven days of the week. The seven celestial objects in table 4.3 (above) were known at the time of Ptolemy and are listed in order of their speed across the ecliptic from slowest to fastest. The origin of the ordering of the days of the week is lost in the mists of time, but here is one plausible explanation that can be traced to writers in Ptolemy's era, and even Chaucer. Not only does the table list the celestial objects from slowest to fastest, but the ordering also corresponds to a progression from the furthest celestial sphere (Saturn) to the nearest (the Moon).

In the astrological interpretation each hour of each day is ruled by a particular celestial object and goes from slowest to fastest. The name of the day is associated with celestial object for the first hour. Accordingly, the first hour of the first day is ruled by Saturn, and we get Saturday. Each of the twenty-four hours gets associated to the celestial objects in that order, so we get Saturn, then Jupiter, etc. until we get to the Moon. Then it starts all over until we get to the end of the twenty-four-hour day. The first day ends in Mars. At that point the next celestial object, the Sun, rules the first hour of the next day, so we have Sunday after Saturday. This process continues until the seven celestial objects are all named in the first hour of each day of the week. This order of progression outlined above yields the following sequence for the planets representing the first hour of each day:

Saturn
The Sun
The Moon
Mars
Mercury
Jupiter
Venus

In English, the first three days of the week directly reflect the seven celestial objects, Saturday, Sunday, and Monday. In Romance languages, the remaining four days of the week retain the planetary order above: Mars, Mercury, Jupiter, and Venus. For example, in French these would be mardi, mercredi, jeudi, and vendredi. In English, the remaining four days are associated to the same celestial objects as in French, but a correspondence between Nordic gods and Roman gods comes into play. So Jupiter, the king of the gods, is associated to the god Thor and so we have Thursday in English. Likewise, Mars is associated with the Nordic god Tiw, hence Tuesday; Mercury with Woden, Wednesday; and Venus with Fria, Friday.

BEYOND SATURN

The discovery of the planets beyond Saturn gives another fascinating example of how new visions of space become reflected in cultural manifestations. The invention of the telescope and evolution into progressively more powerful instruments fundamentally changed our views of space and allowed us to see into the outer reaches of the solar system. Uranus was discovered in 1781 by astronomer William Herschel. Neptune was first described as a planet in 1846, and Pluto in 1930.

Now there is some controversy surrounding the naming of Pluto as a planet or a dwarf planet, but I'll sidestep this question. Uranus, Neptune, and Pluto were embraced into astrology and given characteristics in the same tradition as the seven celestial objects known in the era of Ptolemy. Common modern associations to these planets are:

Uranus—rebellion, upheaval, technology
Neptune—mystical
Pluto—transformation, destruction, renewal

The names of these planets were typically given by the astronomers who discovered them. Urbain Le Verrier, one of the astronomers who discovered Neptune, gave it its name, continuing in the tradition of naming planets after Greco-Roman gods. Pisces is associated with the native sign of Neptune, God of the sea. Pluto is associated with Scorpio, and Uranus with Aquarius.

AGES

As mentioned above, the Zodiac signs don't align with the present-day constellations that go with the same names. This is due to the wobble of the Earth's rotational axis, sometimes called *equinoctial precession*. This was known at the time of Ptolemy and was discovered by an astronomer, Hipparchus (190–120 BC). A complete cycle of precession takes 26,000 years.

While the Zodiac signs and their constellations lined up at the time of Ptolemy, they have since moved. Currently the astrological boundary between Pisces and Aries is physically in the constellation of Pisces and is moving toward Aquarius. As they say in the coding industry, this isn't a bug, it's a feature. The meaning of this precession is associated with long historical periods. A 26,000-year period implies that the Aries–Pisces Zodiac boundary will transit through one Zodiac sign in approximately 2,600 years.

Prior to the time of Ptolemy, this boundary was in the physical constellation of Aries and the world was in the "Age of Aries."

Given the nature of Aries and its association with Mars, this was labeled as a period when warfare was common. After leaving Aries, the boundary entered into the constellation of Pisces, the fish. Given the association of Jesus with fish, it was thought to signal the onset of a Christian era.

Now, astrologers believe that the transition into each new age occurs well before the Aries–Pisces astrological boundary physically enters the next constellation. There is some debate on how much in advance this happens, but depending on how this is calculated, the general belief is that we are headed into the Age of Aquarius. Perhaps the strongest cultural association with the Age of Aquarius is associated with the hippie culture of the late 1960s and early 1970s and the song of the same name that appears in the 1968 Broadway musical *Hair*.

HOUSES

Another concept in astrology is called houses. It turns out that this feature of astrology has tremendous practical and historic significance in the development of celestial navigation, which I will get to shortly. But, first, what are the houses?

A branch of astrology called *natal astrology* is an astrological forecast based on the alignment of the Zodiac signs and celestial objects at moment and place of your birth. For example, the "Sun sign" is the Zodiac sign the Sun occupies when you are born. The concept of houses is part of natal astrology. At the moment and location of your birth, there is a spatial construction about the sky that is highly egocentric. The sky is divided into twelve houses, each one signifying some aspect of your life. The eastern horizon represents the origin and boundary between houses 12 (above) and 1 (below). Houses 1–6 progress in 30-degree swaths below the horizon past the *nadir* (lowest point) and on to the western horizon, which is the boundary between houses 6 and 7. Houses

House Divisions

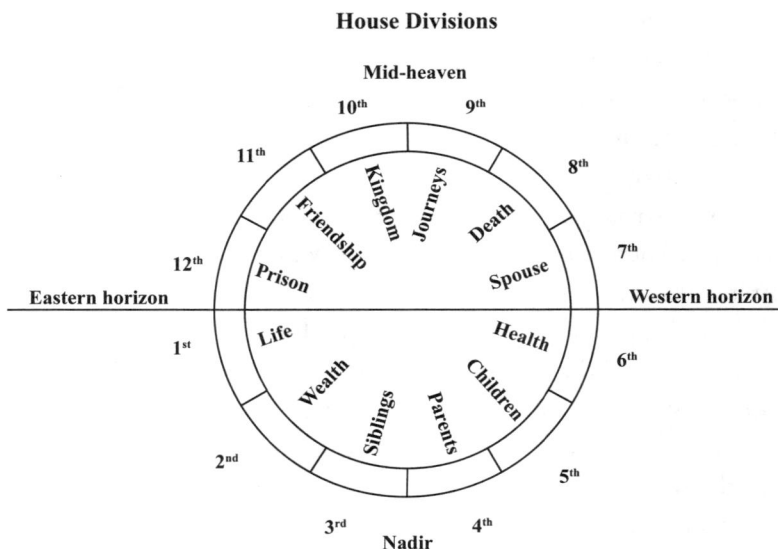

FIGURE 4.3 The construction of the house system with their associations.

7 through 12 proceed upward from the western horizon past the highest point, the *mid-heaven*, and back to the eastern horizon. Figure 4.3 shows the construction and the human connections signified by each house.

One example of a house is the kingdom house, near the mid-heaven. This is an indicator of the nature of a person's vocation. The spouse house represents marriage or a business partner.

At the moment of birth, the signs of the Zodiac and the celestial objects will map into the houses. This will then be a way of foretelling aspects of a person's life. For example, if Mercury is in the kingdom house, this might be an indicator that the person would be a communicator, such as a marketer, as Mercury represents communication. There would be "bonus points" if Virgo, the native house of Mercury, also aligns with the kingdom house. A person might have Mars in their spouse house. Having the planet symbolizing conflict and aggression in the seventh house probably wouldn't bode well for a marriage.

TOOLS OF THE TRADE

Astrology was practiced throughout the Muslim Caliphates in the Middle Ages and beyond. The techniques, which originated with Ptolemy, were continuously updated, and spread to the Western world. As a practical matter, how did astrologers carry out their trade?

First, the astrologers had to have an almanac at their disposal that was current for the period they were making predictions. As such, there had to be practitioners who could use Ptolemy's system of calculating the positions of celestial objects and then publish the results. The predictions in the almanacs give the positions of the Sun, Moon, and planets as a number of degrees into the sign of the Zodiac it inhabits at a specific date.

Second, the astrologers used a table of latitudes and longitudes for locations on Earth to construct the house system for the location of the subject's birth. Like the Almanacs, these were widely available treatises that carried locations in a tabular format. I've done comparisons of modern values of locations for a listing from the eleventh century called the Toledo Tables and found a good agreement for modern accepted values of latitude and reasonable agreement with longitude.

Finally, the astrologers needed to carry out the calculations that gave the positions of celestial object mapped into the houses for the location and time of birth of the subject. This used trigonometric functions.

One commonly used instrument for the astrologers was called an *astrolabe*, which might be considered the equivalent of a modern-day cellphone app with several features (fig. 4.4). It functioned as a star chart, it allowed the user to find the location of the Sun in the Zodiac, and many had a graphical means of extracting the trigonometric functions. The trigonometric functions were necessary to do the calculations to figure out where in the sky a celestial object can be found, given the position in the Zodiac, and the latitude and longitude of the location on Earth.

FIGURE 4.4 Front plate of an astrolabe. This one is on display at the National Archaeological Museum, Madrid, Spain, and dates from 1067 AD. *Photo credit*: Luis Garcia, modified to black-and-white by the author.

CELESTIAL NAVIGATION

One highly consequential spin-off from the practice of astrology is the advent of celestial navigation from the fifteenth century onward. Prior to the fifteenth century, the Arab Caliphates had a monopoly on spice trade from the modern-day Maluku Islands, known as the Spice Islands. In one route they sailed from the Spice Islands up through the

Red Sea, then overland through Egypt, eventually reaching Venice. Seeking to circumvent the Arab monopoly, the Portuguese set out to voyage around the coast of Africa, into the Indian Ocean, establishing bases along the way. Ultimately the Portuguese reached the Spice Islands in 1512.

A major issue in the exploration of the African coast and beyond was the establishment of posts for resupply along the way. Nominally, navigation in the well-known regions of Europe and the Mediterranean was carried out with a compass and observations of Polaris. But, as the Portuguese pushed further south, the sailors eventually lost sight of Polaris.

Here is where astrology came into play. Recall that natal astrology requires a latitude and longitude and a calculation of the position of celestial objects to form a star chart in the house system. By reverse engineering of the calculations for the star chart, by *observing* the locations of stars, planets, the Sun, and Moon, the practice of astrology can be turned around to figure out the position of the observer. In a sense this is a transition from a human-centric view to a more allocentric perspective. The Portuguese placed astrologers on the sailing vessels that departed for voyages along the African coast to do the reverse-engineered observations to find the locations of the supply posts established along the coastline and into India.

Christopher Columbus and his brother Bartholomew worked as mapmakers and sailors in Lisbon from 1477 to 1485 and had the opportunity to observe the practices of the Portuguese navigators. It's there that they conceived of the Enterprise of the Indies: the concept that by sailing west, they could reach the Spice Islands and cut out the arduous trip around the coast of Africa. Eventually they convinced Queen Isabella of Spain to underwrite his famous expedition to the west after being turned down by King John II of Portugal.

Columbus brought the tools of the astrological trade with him and experimented with celestial navigation on his four voyages to the New World. He had an almanac with the positions of the planets, Sun, and Moon, an astrolabe, and a quadrant for making observations.

A quadrant, like an astrolabe, can measure the angle of a celestial object above the horizon. With suitable manipulations and an almanac, these observations can be used to derive a latitude. Columbus's earliest observations were wildly off, but over time his celestial navigation results became more accurate.

Figure 4.5 shows a portion of an almanac that Columbus carried with him on his voyages. It was intended for making astrological charts, but now carried double duty as an aid to celestial navigation. The almanac was produced by mathematician and astrologer Regiomontanus, who used Ptolemy's system to carry out the necessary calculations.

In the entry shown in the figure, the second row shows the symbols for the celestial objects with the Sun, Moon, Saturn, and Jupiter. Below that are the symbols for the locations where they can be found in the Zodiac. In the "Sun" column, you can see that it's in Taurus. The numbers list the position within each Zodiac sign in degrees and arc-minutes. The letter that looks like a *g* is the degrees into Taurus, and the letter that looks like an *m* represents the arc-minute entries. One arc-minute is 1/60th of a degree. As you look down the Sun column, you can see that it is moving roughly one degree per day. On May 1 it's at 20 degrees, 9 arc-minutes in Taurus, and on May 2 it's at 21 degrees, 6 arc-minutes. The Moon moves much more swiftly, and the first entry for a location in Capricorn, but then changes to Aquarius, and then to Pisces from May 1 through May 5. Saturn, also in Capricorn, moves very slowly, at about three arc-minutes per day.

After the first attempts by the Portuguese and Columbus, celestial navigation evolved into a sophisticated practice on most sailing vessels, both commercial and naval. By the end of the nineteenth century, navigators could find their positions with accuracy within a few miles.

The development of celestial navigation had huge consequences. One could argue that it was one of the major factors leading to the hegemony of Western European countries over a large swath of the globe. With the ability to transmit power just about anywhere in

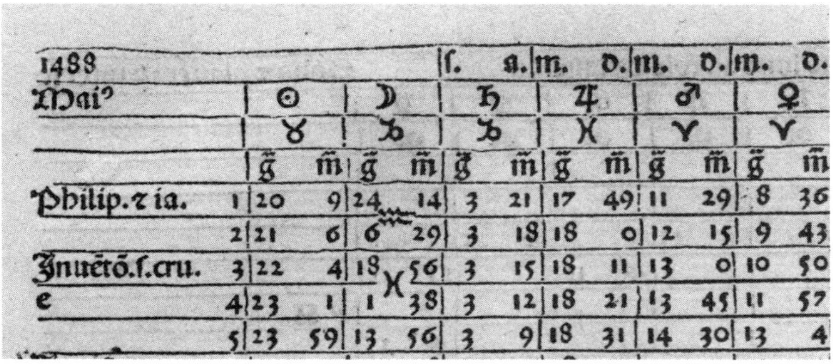

1488				f.	a.	m.	o.	m.	o.	m.	o.
Mai⁹	☉	☽	♄	♃	♂	♀					
	♉	♑	♑	♓	♈	♈					
	g̅	m̅	g̅	m̅	g̅	m̅	g̅	m̅	g̅	m̅	g̅	m̅
Philip. ⁊ ia.	1 20	9 24 14	3 21	17 49	11 29	8 36
	2 21	6 6 29	3 18	18 0	12 15	9 43
Jnuctō.f.cru.	3 22	4 18 56	3 15	18 11	13 0	10 50
e	4 23	1 1 38	3 12	18 21	13 45	11 57
	5 23	59 13 56	3 9	18 31	14 30	13 4

FIGURE 4.5 Section of the almanac developed by Regiomontanus, used by Christopher Columbus on his voyages to the New World. See text for a description. Image courtesy of Harvard University Libraries.

the planet, the major colonizers had an unheralded advantage over other cultures.

An emerging theme is that imagination stirs new visions of space, and the new visions of space become ever more expansive over time. The advent of celestial navigation from astrology illustrates interplay between the human realm and our understanding of space.

Toward the end of the thirteenth century, knowledge of astronomy and astrology was widespread in the cities of Western Europe. Perhaps one of the biggest journeys of the imagination in this era was Dante Alighieri's three-volume *Divine Comedy*, where he constructed a voyage through the center of the Earth, up Mount Purgatory, and into the heavens in Aristotle's model. Throughout the three books, Dante employs his own brand of celestial navigation to mark both time and place as he makes his way through a moral projection onto the universe of his time. I explore this in the next chapter.

* 5 *

DANTE'S JOURNEY

The *Divine Comedy* presents a sustained spatial-social parallel. Dante Alighieri was fascinated with astronomy. In his encyclopedic work *Convivio* he writes of astronomy as being the most "high and noble" of the sciences, with a detailed description of the universe as understood by scholars of that era. As such, the *Commedia* gives a unique poetic window into heavens as seen through his eyes. Sins and virtues are projected onto a space of Dante's imagination that weaves together classical references of the underworld, his invention of Mount Purgatory, and an Aristotelian vision of the sky. In his journey through this universe, Dante makes references to astronomical phenomena to tag locations and note the passage of time in what is effectively an early practice of celestial navigation. He often refers to Zodiac signs to mark progress along his journey.

The *Divine Comedy* is something of a high-water mark of the Aristotelian universe. The Catholic Church populated the celestial spheres with a hierarchy of angels. Here Dante projects human characteristics throughout. Three centuries later the Copernican model and the invention of the telescope changed all of that. Some have seen the *Divine Comedy* as a cultural pivot point. Author Christian Blauvelt writes admiringly:

The *Divine Comedy* is a fulcrum in Western history. It brings together literary and theological expression, pagan and Christian, that came before it while also containing the DNA of the modern world to come. It may not hold the meaning of life, but it is Western literature's very own theory of everything.[1]

In this chapter I focus mostly on Dante's astronomical and geographical references and their context in his journey. Here I owe a big nod of gratitude to Mary Ackworth Evershed (aka Mary Orr) who wrote *Dante and the Early Astronomers*, which details a lot of the sources of Dante's astronomical references in addition to illuminating passages in the *Divine Comedy* through the lens of geography. In the *Divine Comedy*, there are nearly a hundred references to celestial events. Rather than be encyclopedic, I present a limited subset to give a flavor of how Dante incorporates these as space and time indicators.

DANTE'S UNIVERSE

A common representation of the world in the Middle Ages was a T-O map, illustrated in figure 5.1. The *O* in the map is the encircling ocean, which harkens back to the ancient Greek models, with the top of the *T* formed by the Don and Nile Rivers and the Mediterranean Sea represented by the vertical. The *T* divides land into Asia at the top, Europe on the lower left, and Africa on the lower right. Classically, Jerusalem is in the middle.

As with the ancient Greeks, many scholars in the West during Dante's era also believed that there was an occupied half of the Earth and an unoccupied ocean. This model is a crucial component to the structure of space in the *Divine Comedy*. Figure 5.2, based on several common portrayals, shows Dante's universe, with the unoccupied ocean and Jerusalem at the center of the occupied half. These jointly constitute the elements of Earth and Water. The Earth is surrounded by the sphere of air, followed by the sphere of fire, as in Aristotle's model.

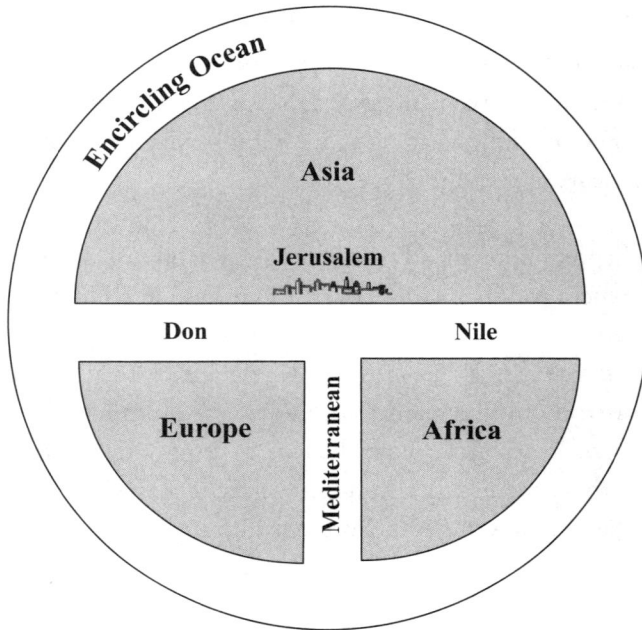

FIGURE 5.1 Illustration of a typical T-O map common in the Middle Ages. The known regions of the Earth to Western Europeans in the time of Dante were depicted as surrounded by an encircling ocean. The vertical of the *T* is the Mediterranean, dividing Europe from Africa. The top of the *T* are the Nile and Don Rivers, dividing Europe and Africa from Asia. Jerusalem is positioned at the center of the map.

Hell is a set of concentric circular ledges that form a kind of staircase to the center of the Earth where Lucifer resides in a frozen lake (fig. 5.3). Past the center is a passageway that leads to the other side of the Earth and ends at Mount Purgatory.

Although Purgatory was a concept in the Catholic Church, the identification of it with a physical mountain on Earth was Dante's invention. He places it at the opposite end of the Earth from Jerusalem, with an Earthly Paradise at its summit. Beyond the summit are the classic Aristotelian spheres starting with the innermost Moon, and the penultimate sphere of Saturn, the Firmament, and the Prime Mover. Beyond this is the Empyrean, the abode of God.

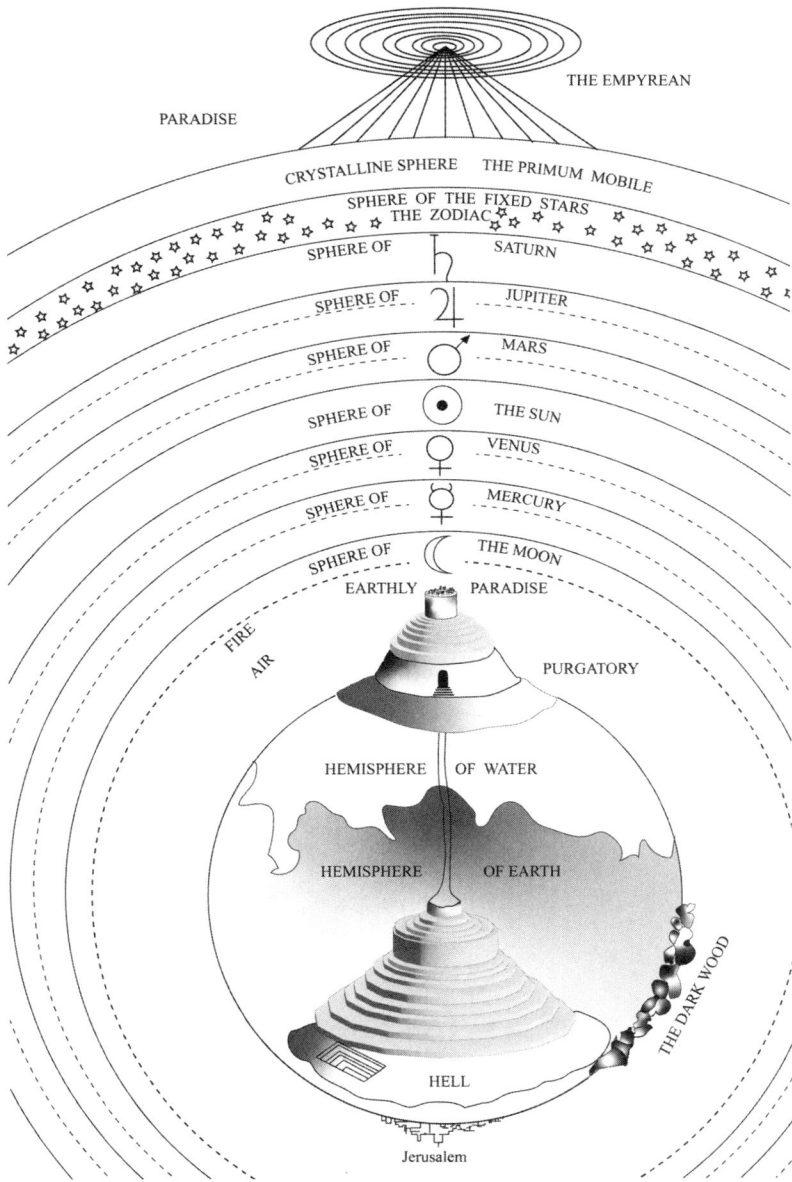

THE EMPYREAN

PARADISE

CRYSTALLINE SPHERE   THE PRIMUM MOBILE

SPHERE OF THE FIXED STARS
THE ZODIAC

SPHERE OF   ♄   SATURN

SPHERE OF   ♃   JUPITER

SPHERE OF   ♂   MARS

SPHERE OF   ☉   THE SUN

SPHERE OF   ♀   VENUS

SPHERE OF   ☿   MERCURY

SPHERE OF   ☽   THE MOON

EARTHLY   PARADISE

FIRE

AIR

PURGATORY

HEMISPHERE   OF WATER

HEMISPHERE   OF EARTH

THE DARK WOOD

HELL

Jerusalem

FIGURE 5.2 Dante's universe mirrors Aristotle's (see fig. 2.6). Earth is
half earth, half water, surrounded by a sphere of air, then a sphere of fire.
Purgatory projects out of the ocean at the antipodal point from Jerusalem
toward the heavens. The heavens themselves consist of the Sun, Moon,
planets, and Firmament. Finally, there is the Prime Mover (Primum Mobile)
that projects motion. Beyond this is the Empyrean, the abode of God.

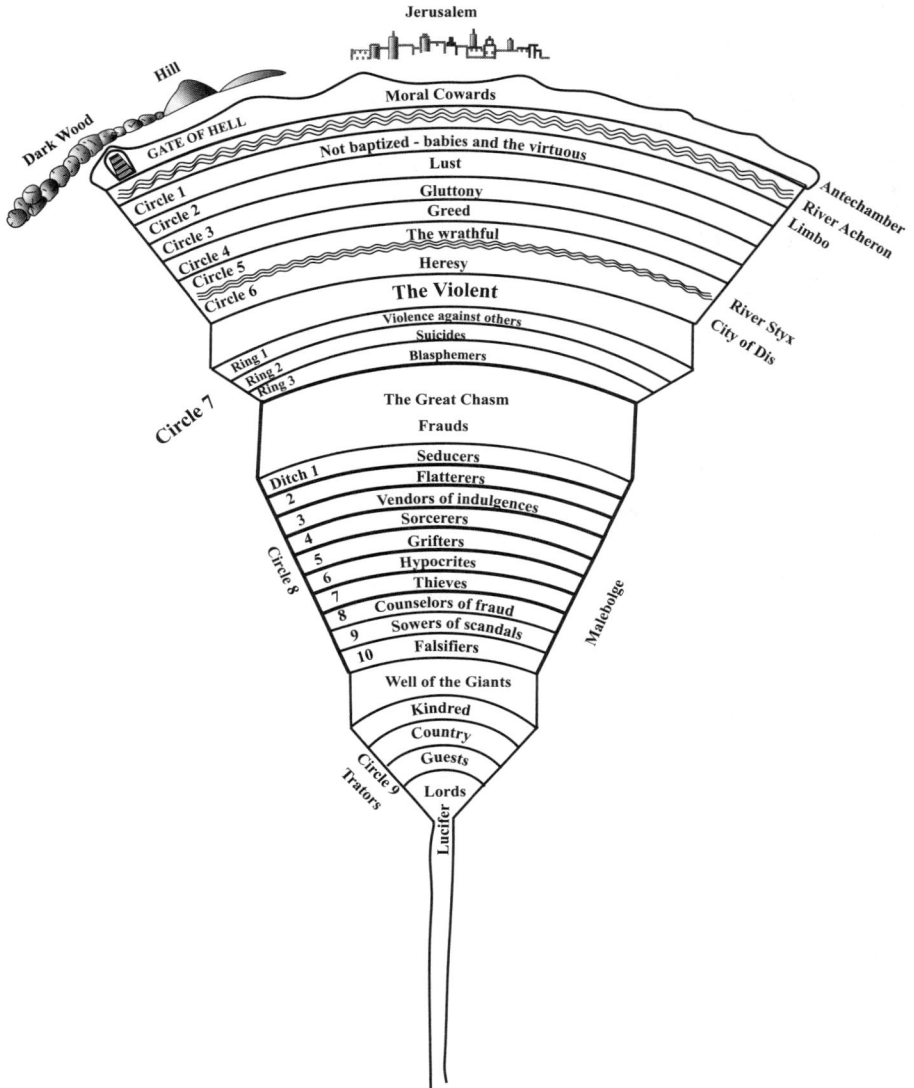

Jerusalem

Moral Cowards

Hill

Dark Wood

GATE OF HELL

Not baptized - babies and the virtuous

Lust

Circle 1

Circle 2   Gluttony

Circle 3   Greed

Circle 4   The wrathful

Circle 5

Circle 6   Heresy

Antechamber
River Acheron
Limbo

The Violent

Violence against others

Suicides

Blasphemers

Ring 1
Ring 2
Ring 3

River Styx
City of Dis

Circle 7

The Great Chasm

Frauds

Seducers

Ditch 1   Flatterers

2   Vendors of indulgences

3   Sorcerers

4   Grifters

5   Hypocrites

6   Thieves

7   Counselors of fraud

8   Sowers of scandals

9

10   Falsifiers

Circle 8

Malebolge

Well of the Giants

Kindred

Country

Guests

Circle 9   Lords

Trators

Lucifer

FIGURE 5.3 Dante's structure of Hell. He and Virgil enter in a portal in the dark woods and cross the River Acheron into the Inferno proper, where sinners occupy a series of concentric circles. The sins are more severe the deeper they penetrate. Below the City of Dis are the Malebolge ditches, representing the eighth circle. The ninth circle is populated by traitors of different varieties. Lucifer resides in the center, and a passageway on the other side leads to the base of Mount Purgatory.

Dante makes references to celestial phenomena throughout the three books: *Inferno*, *Purgatorio*, and *Paradiso*. Each of these three books is broken into a series of cantos, which is a common dividing system for long poetry in the Middle Ages and into the modern era.

To say that Dante was taken with astronomy is an understatement. He wrote extensively about celestial events in his encyclopedic work *Convivio*. The final word in each of the three books in the *Divine Comedy* is "stars" (*stelle*).

INFERNO

The circles of Hell, in order of increasing severity, are:

1. Limbo
2. Lust
3. Gluttony
4. Greed
5. Wrath
6. Heresy
7. Violence
8. Fraud
9. Treachery

Hell contains a number of imaginary geographical features such as the River Styx (Wrath), the City of Dis (Heresy). In addition, there is substructure to some of the circles, particularly in lower Hell.

In the opening scene of *Inferno*, Dante finds himself lost in a dark wood. The first stanza has a perfect spatial/temporal/social metaphor where Dante speaks of himself. I'll try to use mostly English in translation, but the opening stanza is so beautiful in the original that I'll put it here:

Nel mezzo del cammin di nostra vita
Mi ritrovai per una selva oscura
Che la diritta via era smarrita

(*Inf.* 1 1–3)

Here is my rough translation, with care to emphasize the spatial-social parallels:

Midway through the road of our life
I found myself in a dark wood
For the direct path was lost[2]

There is the physical sense of Dante being lost, unable to regain his bearings. In a moral sense he is lost in a kind of mid-life crisis. Dante suggests that he entered the woods almost as a sleepwalker, not really understanding how he got there. He finds himself at the base of a large hill. He gazes on the hill and sees sunlight illuminating its slope. Dante begins to climb the hill but spots a leopard and turns back.

Here is where Dante makes his first reference to celestial events. In this case, he's marking the time of the year as spring (I use the translations by Allen Mandelbaum, unless otherwise noted[3]):

The time was the beginning of the morning;
the sun was rising now in fellowship
with the same stars that had escorted it

when Divine Love first moved those things of beauty;

(*Inf.* 1 37–40)

The sign of Aries, being the first in the Zodiac, is associated with creation and springtime. So, "when Divine Love first moves those things of beauty" is the beginning associated with Aries. Dante is fixing the time as being dawn with the Sun being in Aries. In our calendar this is from March 21, the first day of Spring, through April 19. Further along, in canto 21 we learn that this is Good Friday.

Dante sees a lion, and then a she-wolf, whereupon he loses all hope of climbing the hill. Traditionally the leopard is associated with the sin of lust, the lion with pride, and the wolf with greed.

Virgil, the author of the *Aeneid* and a hero of Dante's, appears. Virgil says that he cannot escape the dark woods by climbing the hill, as he would be killed by the she-wolf. Dante must follow Virgil through an "eternal place," and they set out.

As the day wanes, Virgil explains that he was sent to rescue Dante by a woman from Heaven named Beatrice. There was a real-life Beatrice whom Dante knew in Florence, but she died at a relatively young age. Some would say smitten in love, Dante wrote extensively of Beatrice in his earlier work *La Vita Nuova*. In Paradiso she becomes Dante's guide through Heaven.

At nightfall Virgil and Dante reach the gates of Hell proper, seeing the famous inscription atop the gates:

> Before me nothing but eternal things
> were made, and I endure eternally.
> Abandon all hope ye who enter here.

(*Inf.* 3 7–9)

They enter Hell at sunset.

In a kind of antechamber before Hell proper, Dante and Virgil pass the neutral or moral cowards, who are neither devoted to love nor evil, and are stung over and over by horseflies and wasps. They curse God, their parents, and the time and place of their birth.

President John Kennedy worked this concept of moral cowardice into speeches. At a speech in Bonn, Germany, he declared, "Dante once said that the hottest places in hell are reserved for those who in periods of moral crisis maintain their neutrality."[4] Dante never said this, and his structure of Hell places the moral cowards in the antechamber. The hottest places are reserved for other sins, like the fraudulent or traitors.

On their journey, Dante and Virgil next reach a significant geographic feature: the river Acheron. Souls are brought across the river by ferryman Charon. This is a common reference as the entrance to the underworld in Greek mythology. Dante populates both the *Inferno* and *Purgatory* with five rivers from Greek mythology, the Acheron being one.

After crossing Acheron, Dante and Virgil enter the first circle of Hell: Limbo. The word "limbo" comes from medieval Latin meaning "edge" or "border," and this is very much an edge of sorts. The Catholic Church's Limbo is for babies who die before they are baptized. Another use of Limbo was for Old Testament notables who died before Christ. In an act known as the Harrowing of Hell, in the period between his death on the cross and the resurrection, Jesus is said to have descended into the Earth and liberated the Old Testament worthy, such as Noah and Moses, to Heaven. In 2007 Pope Benedict produced a document that expressed hope that God would save unbaptized infants. This is widely seen as "canceling" Limbo.

Dante goes beyond the classic version of Limbo and populates it with non-Christian souls he sees as worthy. Walking through a wooded area, Dante and Virgil see the shades of various notables from antiquity: Homer, Horace, Ovid, and Lucan. Virgil himself is consigned to Limbo.

They reach a castle with a moat around it and seven portals and meet the souls of ancient Greek thinkers, including Thales, Democritus, Euclid, Ptolemy, Socrates, and Plato. Dante does not mention Aristotle by name but writes of him as a kind of master figure among the rest.

Virgil and Dante then descend through the circles of the sinners in upper Hell: Lust, Gluttony, and Greed. Dante notes the time as midnight as they cross out of the fourth circle of Hell:

> But let us descend to greater sorrow,
> for every star that rose when I first moved
> is setting now; we cannot stay too long

(*Inf.* 7 97–99)

This is a reference to the motion of stars. We can divide the sky for the observer into two halves. There is an imaginary line called the *meridian* that runs from north through the highest point in the sky, the zenith, to south. All stars to the east of that line are rising, and all stars to the west of that line are setting. When Dante and Virgil entered Hell, it was sunset. There was a set of stars to the east of their meridian at that time, but now they are in the western half, approximately six hours later. This raises the unanswered question of how they could see stars from inside Hell.

The next major geographic feature they encounter is the River Styx, also a major river of the mythological Greek underworld. Dante places the wrathful in the Styx, which is more of a swamp than a river, with a stream feeding it from above. Here is the circle of the wrathful.

> When it has reached the foot of those malign
> gray slopes, that melancholy stream descends,
> forming a swamp that bears the name Styx
>
> And I, who was intent on watching it,
> could make out muddied people in that slime,
> all naked and their faces furious
>
> (*Inf.* 7 106–11)

On the far side of the River Styx, they arrive at the City of Dis, a walled city with gates and ramparts guarded by furies and Medusa. Dante spots mosques glowing with fire, as he assigns heretics, including Muslims, to this sixth circle.

The name of the city Dis comes from ancient Roman Father Dis, who is the keeper of Hades, and appears in Virgil's *Aeneid*. It has iron walls guarded by fallen angels and furies and has the features of a guarded city that would be familiar in Dante's era with towers, gates, moats, and bridges. "Dis" itself is derived from "deus" (god). The god of the underworld, "Father Death," was "Dis Pater" in Latin.

The City of Dis is a kind of frontier to the lower circles of Hell that includes the violent, frauds, and traitors.

At this juncture Dante wonders how Virgil knows the territory so well, as he (Virgil) was consigned to the outermost circle of Limbo. Virgil says that once before he went to the deepest depths of Hell to rescue a soul in the ninth circle, where Judas was punished, so he knows the pathway well.

They make their way through Dis, and then to the wood of suicides, arriving at the edge of a great cliff. Virgil urges Dante onward. Here, we have another description marking time from the positions of stars. Dante often cloaks his identification of his locations in an indirect and poetic fashion, which requires some decoding on the part of the reader. Virgil says to Dante:

"But follow me, for it is time to move;
the Fishes glitter now on the horizon
all the Wain is spread out over Caurus;

Only beyond, can one climb down the cliff."

(*Inf.* 11 112–15)

Here, the "Fishes" is Pisces. In spring, if Pisces is on the horizon, it is about two hours before sunrise as the Sun is in Aries, which is the next zodiacal sign to rise. This would make it roughly 4 a.m.

"Wain" is a Nordic term for the Big Dipper. Caurus is a direction in the wind-compass system used by the Greeks and Romans. For example, Zephyrus (Greek) is the west wind, and is a compass direction. Caurus (Latin) means the northwest wind and corresponds to the direction northwest. Seen from the mid-latitudes in early spring, this perfectly describes the alignment of the stars around 4 a.m. (local time) as seen from Florence.

The cliff is the next major geographic feature Dante and Virgil encounter. They climb down it, aided by piles of stones that resulted from a massive earthquake that occurred when Christ died on the cross. This event is referenced in the Gospel of Matthew:

> And when Jesus had cried out again in a loud voice, he gave up his spirit. At that moment, the curtain of the temple was torn in two from top to bottom. The earth shook, the rocks split and the tombs broke open. (Matt. 27:50–51)

Dante and Virgil now enter the eighth circle of Hell. This is called the Malebolge, which is divided into ten concentric ditches. A set of berms or bridges run like bicycle spokes from the outermost to the innermost ditch. Dante and Virgil traverse the ditches on top of the berms, witnessing the trapped sinners suffering their punishments inside.

Dante assigns the fourth ditch to diviners in canto 20. He mentions specific astrologers by name, but doesn't really assail astrology, as it was quite popular in Dante's time, although it was still at odds with the Catholic Church.

In canto 21 another time indicator for Dante comes from the mouth of a demon. Here Dante and Virgil pass through the fifth ditch of the Malebolge, reserved for grifters. This is something of a slapstick description of a gang of devils, with gutter humor. Dante and Virgil are confronted by this band of demons and Dante is instructed by Virgil to hide. The head of this gang, Malacoda, comes forth and converses with Virgil, who asks for directions. Malacoda says that the sixth bridge is smashed and gives a time for it:

> Five hours from this hour yesterday,
> one thousand and two hundred sixty-six
> years passed since that roadway was shattered here.
>
> (*Inf.* 21 112–14)

This is another reference to the earthquake at the moment of Christ's death on the cross, which also was responsible for the rockfall at the Great Cliff. This stanza, buried in canto 21, is the construction that puts the opening of the *Commedia* at the start of Good Friday. The precision of 1,266 years, one day, and five hours is impressive for a demon. In his encyclopedic work *Convivio*, Dante puts the death of Christ at 34 AD.[5] Simply doing the math suggests that the year of Dante's journey is 1300. This stanza puts the time of this stanza as the morning of Holy Saturday, consistent with being some hours after Dante sees Pisces on the horizon while atop the Great Cliff.

Scholars over the centuries have tried to assign a precise date to the *Divine Comedy* based on the numerous references to astronomical phenomena. This stanza is a rather crucial point, but then the challenge is to assemble all the nearly one hundred celestial references into a coherent body. This is not possible without encountering contradictions.[6]

Dante has a significant encounter with the spirit of Ulysses in canto 26. This ties into the notion of an occupied half of the world and an unoccupied hemisphere of ocean with Mount Purgatory rising out of the waters.

The meeting with Ulysses takes place in the eighth ditch of frauds. The sinners in this trench appear as speaking tongues of flames. Ulysses and the warrior Diomedes are merged into one flame. Dante wants to talk with them but, since they speak Greek and Dante doesn't, Virgil translates. Their fraud is the Trojan Horse. Virgil asks Ulysses to explain his fate, and Ulysses answers with a description of a voyage that ends at Mount Purgatory. In his day, Dante knew roughly the tale of Ulysses, but since there were no existing Latin translations of the *Odyssey*, he did not know specifics of Homer's version and creates his own story in its stead.

Rather than voyage back home to his wife Penelope, as from Homer, Dante's Ulysses implores his crew to sail beyond the Gates of Hercules, what we now call Gibraltar. Geographically, this was a symbol of the western limit of the inhabited world. Ulysses and crew then steer a southerly course for five lunar months. As Ulysses and crew approach Mount Purgatory, a whirlwind sweeps down on his vessel and crew, sinking it.

Passing from the Malebolge, Dante and Virgil enter the ninth circle, which is a large frozen lake divided into four concentric rings allocated to traitors of various kinds: traitors to family, community, guests, and lords, respectively. The lake is called Cocytus, the name given to one of the rivers in Hades in ancient Greek mythology. Frozen into the middle of Cocytus is Lucifer himself.

There is a significant moment when Dante and Virgil cross the center of the Earth. They physically climb on the body of Lucifer. Virgil asks Dante to grasp on to his neck. They clamber down to a point at Satan's hip and, with a great effort, Virgil flips head-over-heels at the very center. Virgil explains to Dante that they just passed the center of the Earth. He includes an Aristotelean explanation about how "all weights are drawn."

> And you were there as long as I descended;
> but when I turned, that's when you passed the point
> to which, from every part, all weights are drawn.

(*Inf.* 34 109–11)

Dante and Virgil now follow a passageway that leads to Mount Purgatory. Dante notes that, now that they've passed the center of the Earth, they're moving upward. Day has been switched from night. There is also a movement toward the hemisphere of water. They follow this path away from the center, Dante following:

> My guide and I came on that hidden road
> to make our way back into the bright world;
> and with no care for any rest, we climbed—
>
> he first, I following—until I saw,
> through a round opening, some of those things
> of beauty Heaven bears. It was from there
>
> that we emerged, to see—once more—the stars.

(*Inf.* 34 133–39)

As mentioned above, the end of each book of the *Divine Comedy*: *Inferno*, *Purgatorio*, and *Paradiso*, ends with the word "stars"—more precisely, *stelle*.

PURGATORIO

In the *Inferno* Dante borrowed from both the Catholic Church and ancient Greece mythology to create a structure of Hell, with Limbo, with rivers like the Acheron and Styx. Mount Purgatory itself and its location is a creation of Dante's imagination, although the *concept* of Purgatory was a real thing for the Church. On Judgment Day, Purgatory disappears and those detained on the mountain will all ascend to Heaven. As in the *Inferno*, Dante uses celestial sightings to place the location of Mount Purgatory and denote time as he and Virgil make their climb.

The mountain is at the antipodal point of Jerusalem, and we can look at where this would physically be on the globe. In our modern schema that uses the Greenwich Prime Meridian as the origin of longitude, Jerusalem is at 32° N, 35° E (latitude/longitude). This implies that Mount Purgatory would be located at 32° S, 145° W. A look at the globe shows that this is in the middle of a large empty swath of the Pacific Ocean, far from land. The nearest is the island of Rapa Iti at 28° S 144° W, 300 miles away from the antipode of Jerusalem. It is inhabited, and the native language is a variant of the Polynesian language family. In some ways it's remarkable that the antipodal point is indeed in the middle of the vast Pacific.

The name Purgatory is related to "purging"—the idea being that, although a person may be guilty of a sin, they can be purged of their sin and ultimately enter Heaven, unlike those in Hell. Dante writes:

> and what I sing will be that second kingdom,
> in which the human soul is cleansed of sin,
> becoming worthy of ascent to Heaven.

(*Purg.* 1 4–6)

Figure 5.4 shows Mount Purgatory, according to Dante's construct.

Mount Purgatory has an Ante-Purgatory at the base that is the domain of the late repentant. There are two stages to this base: the excommunicated and then the negligent/indolent. Above Ante-Purgatory is Purgatory proper where souls end up who have committed one of the seven deadly sins:

Pride
Envy
Wrath
Sloth
Avarice
Gluttony
Lust

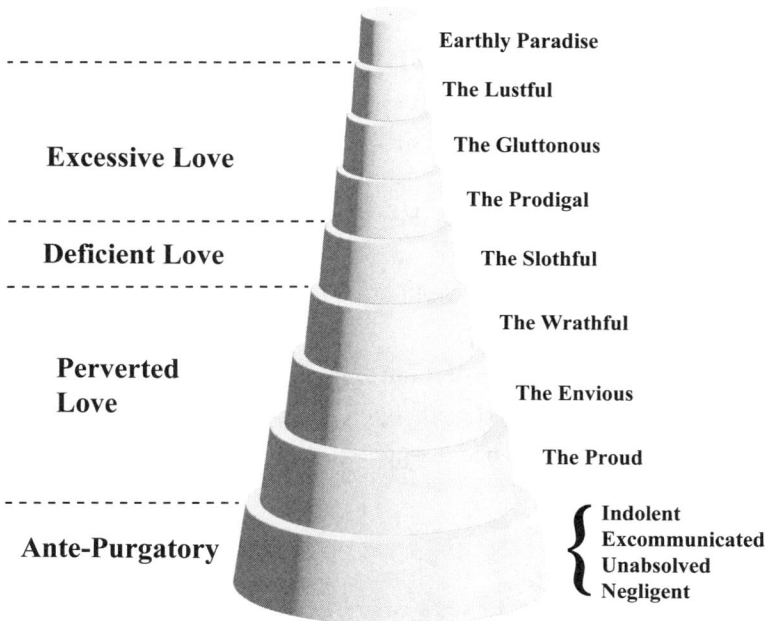

FIGURE 5.4 The structure of Mount Purgatory, with an Ante-Purgatory and a series of terraces where the souls are located. Ante-Purgatory, being at the base, is closest to Hell, and the Earthly Paradise at the top is closest to Heaven.

When Dante and Virgil arrive at the foot of the mountain, it is before dawn, with Venus and Pisces rising:

> The lovely planet that is patroness
> of love made all the eastern heavens glad,
> veiling the Pisces in the train she led.
>
> (*Purg.* 1 19–20)

Recall in the chapter on planets, Venus is never seen far from the Sun, so sighting Venus in the predawn hours is consistent with that positioning. Recall, too, that Pisces was rising in the east about two hours before the Sun when Dante and Virgil were at the top of the Great Cliff.

Dante turns to his right. If he'd been facing east toward rising Venus, he would now be facing south.

> Then I turned to the right, setting my mind
> upon the other pole, and saw four stars
> not seen before except by the first people.
>
> Heaven appeared to revel in their flames:
> o northern hemisphere, because you were
> denied that sight, you are a widower!
>
> After my eyes took leave of those four stars,
> turning a little toward the other pole,
> from which the Wain had disappeared by now,
>
> (*Purg.* 1 22–30)

We don't know what four stars he's referring to. In fact they are said to have symbolic content, but, given the precision of other mentions of stars and planets, it's worth entertaining what they may have represented. If they had not been seen except by the "first people," they would not have names. The "first people" is probably a reference to the Garden of Eden, which is the Earthly Paradise at the top of Mount Purgatory in Dante's formulation.

It's tempting to speculate that the four stars were the stars of the Southern Cross. The Southern Cross is visible from as far north as parts of the Northern Hemisphere, with the southernmost star, Acrux, visible as far north as 26° N latitude. Ptolemy cataloged forty-eight constellations, including many in the southern half of the sky, including Centaurus, which is very close to the Southern Cross, so it's a bit of a stretch to imagine these four stars were the Southern Cross; more likely, they were something Dante made up as part of his narrative. He then turns to look northward, and cannot see Wain, the Big Dipper, which is fully below the horizon at that time from the latitude of Purgatory.

THE SUN'S PATH THROUGH THE SKY

Dante and Virgil walk along a plane and arrive at the shoreline. At this point Dante invokes a kind of celestial navigation to designate both time and the location of Mount Purgatory. He does this at three points in *Purgatorio*. The first time is in canto 2:

> By now the sun was crossing the horizon
> of the meridian whose highest point
> covers Jerusalem; and from the Ganges,
>
> night, circling opposite the sun, was moving
> together with the Scales that, when the length
> of dark defeats the day, desert night's hands;
>
> so that, above the shore that I had reached,
> the fair Aurora's white and scarlet cheeks
> were, as Aurora aged, becoming orange.

(*Purg.* 2 1–9)

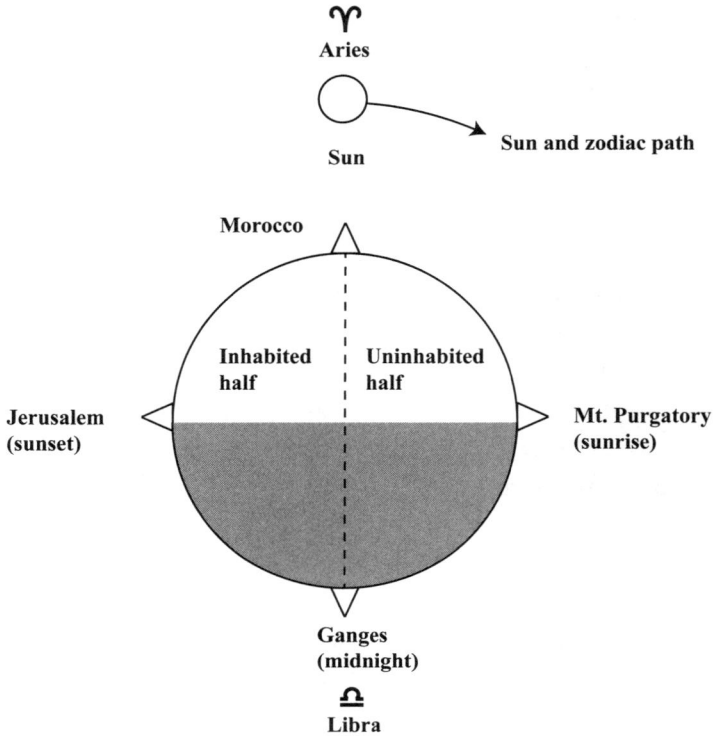

FIGURE 5.5 Diagram of the position of the Sun in the sky at the beginning of canto 2, with Mount Purgatory at dawn. This diagram is looking down on the North Pole. The Sun is at noon over Morocco, and Jerusalem is at sunset.

If you drew a complete circle of a line of longitude (a meridian) from Purgatory through the North Pole and to the other side, Jerusalem would lie on the continuation of that meridian line. In Dante's construct, he places Morocco as the westernmost part of the inhabited land, 90° of longitude away from Jerusalem, and Ganges (the mouth of the Ganges) as a kind of marker for the easternmost part of the inhabited land, 90° away from Jerusalem as well. Morocco, Cadiz Spain, and the Straits of Gibraltar are all used to denote the westernmost point of land in various places in the *Divine Comedy*.

Refer to figure 5.5 as a decoding of these stanzas. When the Sun is rising at Mount Purgatory, it's just setting in Jerusalem. Then Ganges would be at midnight. The Sun is still in Aries, as only a few days have elapsed since the opening. Libra is the sign opposite to Aries in the Zodiac, so that the "Scales" would coincide with Ganges at that moment.

This is all a kind of mashup of both time and location that allows Dante to place Mount Purgatory at the antipode of Jerusalem explicitly. This kind of association of celestial objects with respect to time and lines of longitude is the basis, in part, for celestial navigation. Here Dante himself is using the positions of celestial objects in the sky to give both time and position.

A similar construction is used later. In canto 4 they climb up the very steep face of Mount Purgatory, which gets easier as they ascend. Virgil comments on the time that elapsed since they started climbing:

> And now the poet climbed ahead, before me,
> and said: "It's time; see the meridian
> touched by the sun; elsewhere, along the Ocean,
>
> night now has set its foot upon Morocco."
>
> (*Purg.* 3 137–39)

Here I take the "meridian" to mean the local meridian, meaning that the Sun is at its highest point in the sky, and using Dante's construct for locating Purgatory this implies that Morocco is now experiencing sunset (fig. 5.6).

I'll jump way ahead to the time when Dante nears the top of Mount Purgatory and must traverse the zone of fire. This is approaching the end of *Purgatorio*. Recall the sublunary world is divided into the element Earth, represented by the inhabited part of the world, and the interior, then the element Water—the uninhabited part. Mount Purgatory thrusts into the element Air, and then the final sphere is the element Fire, which Dante must traverse to reach the Earthly Paradise. Before he makes the traverse, the sun sets:

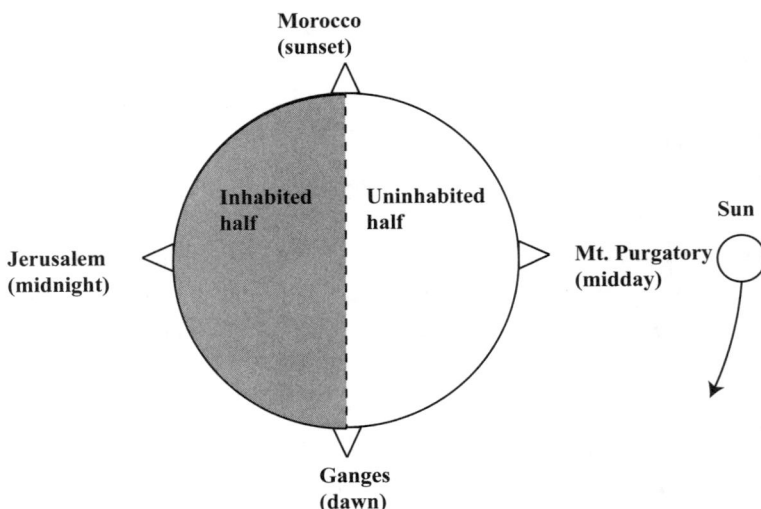

FIGURE 5.6 The relative positioning of the Sun in
canto 4 at midday on Mount Purgatory.

Just as, there where its maker shed His blood,
The sun shed its first rays, and Ebro lay
Beneath high Libra, and the ninth hour's rays

Were scorching Ganges' waves; so here, the sun
Stood at the point of day's departure when
God's angel—happy—showed himself to us.

(*Purg.* 27 1–6)

Again, with the now familiar diagram, you have first sunrise at
Jerusalem. The Ebro River is in Spain, so this corresponds to the
Morocco/Spain location as the westernmost point of land. Libra is
opposite Aries and the Sun. The "ninth hour" is *nona*, which is the
root of noon. By the twelfth century it became associated with midday. The Sun is "scorching" Ganges, which is presumably the local
noon. Finally, Mount Purgatory is at the "point of day's departure"

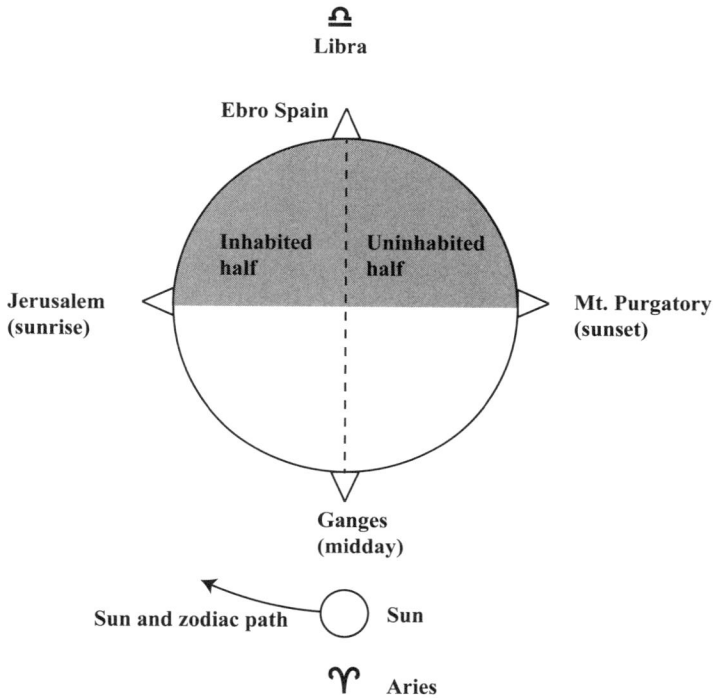

FIGURE 5.7 Position of the Sun in the sky relative to Libra, Aries, and geographic locations on Earth. This is from canto 27.

(fig. 5.7). "God's angel" refers to the angel of chastity who appears on the terrace with them.

Dante and Virgil pass into the Earthly Paradise. Virgil departs, and a mysterious shrouded woman appears, and says,

> "Look here! For I am Beatrice, I am!
> How were you able to ascend the mountain?
> did you not know that man is happy here?"

(*Purg.* 30 73–75)

Dante lowers his eyes and sees a stream but is still somewhat in shame. Beatrice bemoans the fact that Dante could not help himself despite her appearance in his dreams. She laments that Dante's journey through Hell and Purgatory was the only path to his salvation:

> He fell so far there were no other means
> to lead him to salvation, except this:
> to let him see the people who were lost.
>
> For this I visited the gateway of
> the dead; to him who guided him above
> my prayers were offered even as I wept.
>
> (*Purg.* 30 136–42)

Here we encounter the last two of five rivers from ancient Greek mythology appearing in the *Divine Comedy*: the rivers Lethe and Eunoe. In crossing the River Lethe, souls forget their mortal sins. In crossing Eunoe, the memories of good deeds in life are strengthened. In the last canto of Purgatorio, Beatrice leads him to Eunoe. After drinking from the Eunoe, he feels renewed and, "I was pure and prepared to climb unto the stars."

Again, *stelle*.

PARADISO

A more challenging book than *Inferno* and *Purgatorio*, *Paradiso* has more of a philosophical bent to it, and even Dante warns the reader of the challenges in proceeding at the opening.

The structure of Paradise is that of Aristotle's universe with each of the spheres being a virtue:

Moon—Inconstant of vows
Mercury—Lovers of glory
Venus—Lovers
Sun—Theologians

Mars—Martyrs and crusaders
Jupiter—Righteous rulers
Saturn—Contemplatives
Firmament—Triumph of Christ
Primum Mobile (Prime Mover)—Orders of Angels

Finally, there is the Empyrean, the abode of God. After traversing the spheres of the planets and Sun, Dante crosses the Firmament and the Prime Mover, finally arriving at the Empyrean. Here he is outside of space and time, much as in Aristotle's conception of what lies outside the Prime Mover, if *outside* is even a meaningful descriptor.

Rather than enter a lengthy description of Dante and Beatrice's passage through Paradise, I focus here on one point: the perfection or lack thereof for the heavens in the time of Dante. The question of celestial incorruptibility persisted from the time of Aristotle until the seventeenth century, when the invention of the telescope revealed structure on the planets. While the planets appeared as points of light in the fourteenth century, the Moon clearly had a mottled appearance to the naked eye. This aspect of the Moon presented a challenge to the notion that there was perfection in the heavens.

In canto 2 Beatrice and Dante enter the sphere of the Moon, populated by the inconstant in their vows. Dante asks Beatrice how the Moon can show its markings if it's part of a perfect Heaven. Beatrice replies:

> she said: "If the opinion mortals hold
> falls into error when the senses' key
> cannot unlock the truth, you should not be
>
> struck by the arrows of amazement once
> you recognize that reason, even when
> supported by the senses, has short wings."

(*Par.* 2 52–57)

Beyond the Moon, this is a meta-statement about our ability to grasp concepts purely through reason. Dante, through Beatrice, is suggesting that there are truths beyond pure reason.

Dante was perplexed by the mottled surface of the Moon and took it up in his semi-encyclopedic work *Convivio*, which has considerable writings on astronomy. Beatrice answers him by saying that differences can even be seen in the eighth sphere, the firmament, as the stars have different colors and brightness. The differentiations in the firmament, like the force from the Prime Mover, are transmitted downward through the spheres.

Likewise, the eclipse of the Sun shows imperfection. Finally, Beatrice sets up a thought experiment with three mirrors, two on either side of the person, and one at a different distance. Although the images all look different, they are of the same person. This is a kind of the shadows-on-the-cave-wall argument that Plato made.

At the end of *Paradiso*, Dante laments that his intellect cannot solve all his conundrums by reason alone.

> As the geometer intently seeks
> to square the circle, but he cannot reach,
> through thought on thought, the principle he needs,
>
> so I searched that strange sight: I wished to see
> the way in which our human effigy
> suited the circle and found place in it—
>
> and my own wings were far too weak for that.
> But then my mind was struck by light that flashed
> and, with this light, received what it had asked.
>
> Here force failed my high fantasy; but my
> desire and will were moved already—like
> a wheel revolving uniformly—by

the Love that moves the sun and the other stars.

(*Par.* 33 133–45)

Here, we have the final writing of *stelle.*

Three centuries after Dante, the telescope was invented and revealed structure associated with the planets, like the rings of Saturn, phases of Venus, and the moons of Jupiter. Observations triumphed where pure reason could not solve riddles, which spelled the end of the Aristotle's model of space.

But our inclination to project human-like qualities onto space persisted. Where Dante populated the heavens with virtuous souls, some astronomers contemplate whether the universe could be home to intelligent beings like us.

IMAGINING
EXTRATERRESTRIALS

As you've seen by this point in our journey, we're good at projecting onto space. This may be due to the practical rewards of using the heavens to mark the seasons or orienting ourselves at night. This projection has a strong cultural content, as we've seen in astrology and Dante's journey. In the West we have constellations associated with animals like Scorpio and Taurus, or gods like Orion and Cassiopeia. The concept that the skies could also be home to real beings like ourselves gained considerable cultural currency from the nineteenth century onward and can be traced in part to the invention of the telescope.

But first, let's go way back in time to examine belief in other intelligent beings. The concept of a *plurality of worlds* dates from Anaximander (610–546 BC). Recall that Anaximander proposed that the Earth was shaped like a squat cylinder floating in space with the inhabited part on the flat top. He also believed in the possibility of multiple worlds with intelligent inhabitants like humans.

In broad brushstrokes, in ancient Greece there was a long-standing tension between the atomists' and Aristotle's vision of space. Aristotle believed that his fundamental elements of Fire, Water, Earth, and Air were continuous and not discrete. As we've seen, the Aristotelian universe was finite with the firmament and Prime Mover as the outermost spheres. It became heretical to propose that there were other worlds inhabited by beings like us.

In contrast to Aristotle, atomists like Democritus and Epicurus held the universe consisted of discrete units, atoms, that could be arranged and rearranged into different configurations with infinite possibilities. Space was also infinite. With different configurations of atoms possible, their conclusion was that, yes, there would be other worlds out there with intelligent inhabitants.

Roman philosopher Lucretius (roughly 94–51 BC) wrote about atoms and the plurality of worlds and their transience in *De rerum natura* (The nature of things). Lucretius argued in favor of an infinite universe and the existence of other worlds. Meditating on this he drew the conclusion that these worlds can be inhabited by "different peoples" and "wild beasts."

> when an abundant supply of matter is available, when space is at hand and there is no obstruction from any object or force, things most certainly must happen and objects must be created. And if the fund of seeds is so vast that the sum of the lives of all living creatures would not suffice to count it, if the same force of nature is still operative and possesses the power to assemble all the seeds of things in the same order in which they have been assembled in our world, you are bound to admit that in other parts of the universe there are other worlds inhabited by many different peoples and species of wild beasts.[1]

While Lucretius's work was unknown in the Middle Ages, Aristotle's views on the impossibility of other worlds tended to dominate. But in 1417 a manuscript of *De rerum natura* was discovered in a monastery, igniting a revival of the concept of the plurality of worlds.

As mentioned previously, while the Aristotelian model of space was notional, the Ptolemaic model was calculational, although it had many tweaks to make the planetary paths through the skies predictable with some precision. Nicolaus Copernicus (1473–1543), inspired by many who came before, produced a system with the Sun at the center of the universe. Copernicus did not strongly pursue the idea where, for example, a superior almanac might have demonstrated the utility of a heliocentric view. It would take until Johannes Kepler (1571–1630) with his elliptical orbits to demonstrate the utility of a heliocentric model.

Italian philosopher Giordano Bruno (1548–1600) was inspired by both the Copernican model and Lucretius and wrote treatises on the possibility of an infinite universe and other worlds.[2] Bruno was arrested by the Inquisition in 1592 and burned at the stake for heresy in 1600. The reason for his execution is not fully clear, but he espoused a number of anti-Catholic viewpoints that probably led the list, but his views on the plurality of worlds probably didn't help his case.

As regards thinking on extraterrestrial beings, the invention of the telescope became something of a game changer. When Galileo learned of the early work on telescopes, he built his own and turned it to the skies, revealing structure on the planets, which had previously appeared to the naked eye as specks of light: moons orbiting Jupiter, rings around Saturn, and more detail of the Moon, where the circular craters became apparent.

Galileo published his findings in *Siderius nuncius* (Sidereal messenger) in 1610. Kepler speculated that the circular craters on the Moon were constructions of intelligent beings who "make their homes in numerous caves hewn out of that circular embankment."[3] Galileo was more circumspect on such pronouncements, possibly because of Bruno's fate.

Dutch scientist Christiaan Huygens (1629–95) likewise speculated on intelligent life on the planets. Huygens contributed much to the sciences and is particularly known for his work on optics that, in a way, foreshadowed some aspects of modern physics. Concerning astronomy, he studied the rings of Saturn and discovered its largest moon, Titan. He wrote a treatise, *Cosmotheoros*, subtitled (translated from Latin): "Conjectures Concerning the Planetary Worlds, Their Inhabitants and Productions."

The opening of *Cosmotheoros*, published posthumously in 1698, shows the state of play at that time on the knowledge of the solar system and the possibility of life.

> A man that is of Copernicus's opinion, that this earth of ours is a planet, carried round, and enlightened by the sun, like the rest of the planets, cannot but sometimes think, that it is not improbable that the rest of the planets have their dress and furniture, and perhaps, their inhabitants too, as well as this earth of ours: especially, if he considers the later discoveries made in the heavens since Copernicus's time, namely, the attendants of Jupiter and Saturn, and the plane and hilly countries in the moon, which are a strong argument of relation and kin between our earth and them, as well as a proof of the truth of that system.[4]

Huygens pays homage to Kepler as one of the originators of the concept of inhabitants on the Moon. The idea that the planets have a structure (dress and furniture), and possibly inhabitants, presumably comes from the visual views with the telescope.

Since the position of the Catholic Church at the start of the seventeenth century was centrality of the Earth, much effort went into the question of the possible plurality of worlds. Huygens, for example, writes "These conjectures do not contradict the holy scriptures." He writes in *Cosmotheoros*:

> Since then the greatest part of God's creation, that innumerable multitude of stars, is placed out of the reach of any man's eye; and many of them, it is likely, of the best glasses [i.e., telescopes], so that they do not seem to belong to us; is it such an unreasonable opinion to think, that there are some reasonable creatures, who see and admire those glorious bodies at a nearer distance?[5]

Huygens imaginatively populated the visible planets with different beings who he speculated had human-like features, but perhaps modified, like much longer necks or huge eyes.

In the eighteenth and early nineteenth centuries there was considerable debate as to whether the plurality of worlds was consistent with Christian belief or not. Political theorist Thomas Paine, for one, claimed they were incompatible. One line of reasoning asked whether the Messiah could save humans on an uncountable number of other worlds, with all the attendant complications of multiple crucifixions and resurrections.

The concept of a plurality of worlds received a boost from the *nebular hypothesis* proposed by philosopher Immanuel Kant, and more thoroughly developed by mathematician Pierre Laplace. In the hypothesis the solar system came to exist out of a single proto-cloud of matter that then coalesced into the Sun and planets. There are/were several points in favor of this origin story. One is that the planets all move in roughly the same plane, and the orbits are all in the same direction, namely counterclockwise when one looks down toward the North Pole. Although the planetary orbits could be ellipses, they are close to being circular. These views all pointed to a holistic physical explanation of how the solar system came to be. If this is the case here, why would it not then be possible for other solar systems to exist around distant stars? Ultimately, the Kant–Laplace model does not explain some details of the solar system but it was an important step in an understanding of its origin.

The timescale associated with the creation of the solar system was bolstered by geologic evidence and Darwin's theory of evolution. Both considerations point to the long timescales associated with how the Earth and its inhabitants came into being. One particularly famous pluralist, astronomer Camille Flammarion, argued strongly for a plurality of worlds in his work *La pluralité* (1862), based on arguments grounded in the nebular hypothesis, natural selection, and geology.

The question of whether humans and our planet are unique is a reemergence of an allocentric/egocentric divide on a cosmic scale. Arguments were still made that humans are unique, and attempts were made to create models where the solar system held a unique position in the universe. This began to fall apart with the observation of the huge distance scales to spiral galaxies by 1930, however.

MARS IN THE POPULAR IMAGINATION

The dynamic of the visual appearance of planets through a telescope and the imagining of beings on the planets persisted. Mars in particular fed the imagination of astronomers in the nineteenth and twentieth centuries (fig. 6.1). In part the observation of features, including polar ice caps, drove the creative interpretation of details. We get a close look at Mars roughly once every two years when the Earth overtakes it in its orbit at a time called opposition. During these times, astronomers would train their telescopes on Mars to make detailed observations.

During the Martian opposition of 1877 Italian astronomer Giovanni Schiaparelli observed markings on the planet that he called

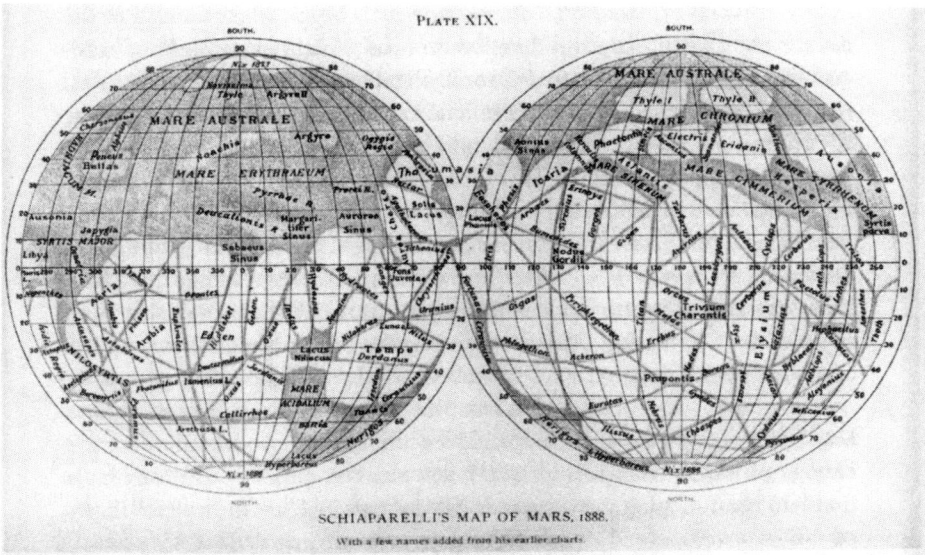

FIGURE 6.1 One of Schiaparelli's drawings of Mars. This was from 1888 and includes the features he called *canali*.

canali, which means "channel" in Italian, often implying a water-course. He believed the polar caps to be made of ice and the melting of the ice caps caused water to flow through the channels. Inspired by Schiaparelli's claim of channels, a few other astronomers reported canals as well as lakes.

Enter Percival Lowell, who took the concept of canals on Mars to dizzying heights. Lowell was born in Boston in 1855 into a wealthy family. After graduating from Harvard, he traveled extensively abroad, and on his return to the United States was inspired by Flammarion to study Mars, with a particular curiosity about Schiaparelli's *canali*. Being independently wealthy, Lowell set out to build an observatory in Flagstaff, Arizona. This was one of the first observatories to be constructed in a location that had excellent viewing conditions—it was high and dry and had little light pollution. On this point alone he was something of a pioneer.

In 1894, another year of opposition with Mars, Lowell trained his telescope on the red planet. Indeed, he wrote of seeing canals. He took a giant step at this point. Perhaps it was the influence of Flammarion, but he claimed not only to see "canals," but that these were constructions of a civilization on Mars. Lowell believed that the purpose of the canals was to take water from the polar ice caps and bring it to the arid regions of the planet to help the Martians survive.

Another development in that era was the offering of the Guzman Prize. Clara Gouget Guzman was a wealthy French widow whose son admired Flammarion's works. When she died, she bequeathed in her will a prize named after her son, Pierre Guzman, the award "to the person of whatever nation who will find the means within the next ten years of communicating with a star (planet or otherwise) and of receiving a response." The prize was announced in 1891 by Flammarion and was to be awarded by the French Académie de Sciences. In the conditions of the prize, communications with Mars were specifically excluded, as it was widely believed that Mars was surely inhabited and communication would be straightforward.

One idea that was publicized in 1892 was a system of mirrors that would reflect flashes of sunlight toward Mars in a system of communications.[6]

Light flashes from Mars were reported in 1892 from the Lick Observatory in California. In the opposition of 1894 there was also an observation of light flashes by astronomers Henri Perrotin and Stephane Javelle from an observatory in Nice, France. This was reported in an issue of *Nature* dated August 2, 1894, "A Strange Light on Mars."[7]

A STRANGE LIGHT ON MARS

Since the arrangements for circulating telegraphic information on astronomical information on astronomical subjects was inaugurated. Dr. Krueger, who is in charge of the Central Bureau at Kiel, certainly has not favored his correspondents with a stranger telegram than the one which he flashed over the world on Monday afternoon:

> "Projection lumineuse dans région austral du terminateur de Mars observée par Javelle 28 Juillet 16 heures Perrotin."

There are rapid response systems of communications among astronomers as a way of reacting to time-sensitive observations. This communication was from the Central Bureau for Astronomical Telegrams, one of the earliest "early warning" systems, that was then picked up on by the journal *Nature*.

The above quote is roughly translated as "Light projection in the southern region of the Mars terminator observed by Javelle July 28 at 4 PM Perrotin." A "terminator" in astronomy is the boundary between the portion of a planet (or Moon) illuminated by the Sun and the night, locally. The editors of *Nature* went on to speculate, "If we assume the light to be on the planet itself, then it must either have a physical or human origin; so it is to be expected that the old idea that the Martians are signalling to us will be revived."

H. G. Wells was an avid reader of *Nature*, and the news appeared as part of the opening chapter in *The War of the Worlds*. Wells himself claimed that part of his inspiration was a discussion with his brother Frank about what would happen if "other beings were to drop out of the sky" and mused over the fate of the Tasmanians when the British invaded. In a way Wells turned the tables on the British empire by imagining an invasion from Mars. He was keenly aware not only of the reports of Perrotin's lights, but also of Lowell's reports of canals. In the serialized debut of *War of the Worlds* in 1897, Wells wrote in the introduction:

> During the opposition of 1894 a great light was seen on the illuminated part of the disk, first at the Lick Observatory, then by Perrotin of Nice, and then by other observers. English readers heard of it first in the issue of *Nature* dated August 2. I am inclined to think that this blaze may have been the casting of the huge gun, in the vast pit sunk into their planet, from which their shots were fired at us. Peculiar markings, as yet unexplained, were seen near the site of that outbreak during the next two oppositions.
>
> The storm burst upon us six years ago now. As Mars approached opposition, Lavelle of Java set the wires of the astronomical exchange palpitating with the amazing intelligence of a huge outbreak of incandescent gas upon the planet. It had occurred towards midnight of the twelfth; and the spectroscope, to which he had at once resorted, indicated a mass of flaming gas, chiefly hydrogen, moving with an enormous velocity towards this earth. This jet of fire had become invisible about a quarter past twelve. He compared it to a colossal puff of flame suddenly and violently squirted out of the planet, "as flaming gases rushed out of a gun."
>
> A singularly appropriate phrase it proved. Yet the next day there was nothing of this in the papers except a little note in the *Daily Telegraph*, and the world went in ignorance of one of the gravest dangers that ever threatened the human race. I might not have heard of the eruption at all had I not met Ogilvy, the well-known astronomer, at Ottershaw. He was immensely excited at the news, and in the excess

of his feelings invited me up to take a turn with him that night in a scrutiny of the red planet.

The reports by the Lick Observatory and Perrotin are mentioned above, and then the remainder is a clever weaving of Wells's imagination, including the use of spectroscopy in astronomy to analyze the nature of the gases. "Lavelle" may have been inspired by Javelle in the original report in *Nature*. Although Wells doesn't mention it explicitly, the speed of the telegraphic astronomical exchange feeds a sense of urgency and speed to the communications. This speed and urgency were also employed by Orson Welles in his radio rendition of the *War of the Worlds* in 1938.

Although there had been science fiction works that invoked extraterrestrials preceding the *War of the Worlds*, this work exploded into the public consciousness and set off a huge cultural revolution. Its cultural impact cannot be overstated, and the trail starting with the invention of the telescope and speculations of astronomers is direct.

It's impossible to catalog all the books, movies, and other works of science fiction that involve extraterrestrials, but even the *War of the Worlds* itself has spawned two movies (1953 and 2005), and the famous radio broadcast by Orson Welles, framed in part as news reporting.

There is an important caution about Lowell's claims. The size of the lenses or mirrors and the wavelength of light combine to place a restriction on how fine-grained an image can be observed. A quick calculation of his telescope's ability to resolve structures shows that Lowell could not have possibly seen as much detail as he represented in his drawings and the canals were likely just products of his overactive imagination. There has been speculation that light from his telescope was illuminating the blood vessels in Lowell's eye, and he was drawing images of his own retina.[8]

In another curious episode Lowell compared his drawings on two successive oppositions and found that they differed significantly. Rather than attribute this to an error on his part, Lowell claimed that the Martians were able to vastly reconfigure their canal system in two years. The "rebuilding" hypothesis was published as a major story in the *New York Times* on August 27, 1911. The headline read, "Martians Build Two Immense Canals in Two Years: Vast engineering works accomplished in an incredibly short time by our planetary neighbors."

In a lecture in 1985 at the University of Glasgow, astronomer Carl Sagan singled out Lowell's advocacy for Martian canals as a cautionary tale of projecting human characteristics onto the cosmos. While history has born this out in Lowell's case, the human tendency to project onto the edge of knowledge is still with us. Sagan is also known for popularizing the phrase, "Extraordinary claims require extraordinary evidence," in reference to reports of extraterrestrial visitations to Earth.

Belief in life on Mars persisted well into the twentieth century. Astronomical technology improved to the point where there could be measurements of surface temperatures, and the composition of the atmosphere, including pressure. As measurements improved, all of these indicated a climate that is quite inhospitable to human life as we know it but didn't rule out the possibility of simpler forms of life.

In 1965 the manmade satellite probe Mariner 4 made a close approach to Mars and relayed photos that didn't show any evidence of canals, but of a partially cratered surface that was reminiscent of the Moon. Since Mariner 4, there have been many visits from Earth, first from flybys, and then landings on Mars. No definitive proof of even primitive life forms has been found. Still, Lowell should be credited in creating intense interest in the possibility of life on Mars.

SEARCHES FOR EXTRATERRESTRIAL INTELLIGENCE (SETI)

Since the publication of *War of the Worlds*, the search for extraterrestrial life has permeated popular culture and likewise obsessed many scientists. However, as of this writing, there has not been any definitive proof of intelligent life outside of Earth, only a few tantalizing possibilities. Here, I'll mostly focus on the possibility of extraterrestrial intelligence, as opposed to simple life forms.

In 1950 Italian physicist Enrico Fermi had a lunch meeting with a few close colleagues at Los Alamos National Laboratory. They got into a discussion about the probability of a visitation by extraterrestrials. Already there had been reported sightings of flying saucers, which none of the scientists believed to be real, but it led to speculations about why there had not yet been conclusive evidence of extraterrestrial life or a direct visit. One recollection was that Fermi asked, "Where are they?" This utterance became known more broadly as the "Fermi paradox," namely, if we believe so strongly in the existence of extraterrestrial life, why have we not found any evidence? Fermi followed up with calculations on the possibility of Earth-like planets out there in the universe and concluded that we should have been visited many times within recorded history.

In 1933 an engineer from Bell Laboratories, Karl Jansky, published a paper reporting on receiving radio waves from an extraterrestrial source, ushering in the era of radio astronomy. Jansky's observation was an accident made when he was trying to track down mysterious static that interfered with transatlantic shortwave-radio communication. By the 1950s radio astronomy was developing the capability of imaging radio sources in space. Moreover, scientists began to view radio telescope as possible ways of listening for communications from advanced extraterrestrial civilizations. In the United States the National Radio Astronomy Observatory (NRAO) was founded in 1956, creating a central facility.

One of the earliest articles on using radio telescopes for SETI was by Giuseppe Cocconi and Phillip Morrison, "Searching for Interstellar Communications," in 1959. They recommended looking at the wavelength of twenty-one centimeters, which is a strong radio emission line in hydrogen, the most plentiful element in the universe. The premise is that it would be a universally understood "best" frequency to broadcast and listen, since it's based on a natural process.

This raises the curious question of what assumptions we make about a civilization out there. Currently, the only model we have is ourselves. Using the constants of nature as a common "language" is probably a good bet, as a developed civilization would hopefully understand science as we do. But this notion could be a horribly egocentric conceit on our part.

At NRAO astronomer Frank Drake made scans at 21 cm of part of the sky in 1960 in Project Ozma, named after Princess Ozma, who appears in L. Frank Baum's Land of Oz series. Baum called Oz "very far away, difficult to reach, and populated by strange and exotic beings."

Drake didn't find any evidence of an extraterrestrial signal in his initial search. Drake then called a meeting at NRAO to discuss a more systematic way of searching for SETI signals. A notable participant was the young Carl Sagan, who became one of the leading proponents of SETI searches, and a face of popular science in the United States.

Drake set out to characterize the probability of finding extraterrestrial life by breaking it down into multiple components in his now famous "Drake Equation." The equation calculates the number of civilizations (N) in our galaxy that have the *capability* of communicating with us. It goes as follows:

$$N = R_* \bullet f_p \bullet n_e \bullet f_l \bullet f_i \bullet f_c \bullet L$$

Where

R_* is the rate of star formation suitable for intelligent life
f_p is the fraction of those stars with planetary systems
n_e is the number of planets in each system with an environment suitable
   for life
f_l is the fraction of suitable planets where life actually appears
f_i is the fraction of life bearing planets on which intelligent life exists
f_c is the fraction of civilizations that have a technology capable of
   broadcasting its existence
L is the average length of time such civilizations produce such
   signs (years).

This was first discussed at a meeting at NRAO in 1961 that included Drake and Sagan. The first three terms in the equation are physical, depending on stars, stars that had planets orbiting, and planets orbiting in a "Goldilocks" zone—not too hot, not too cold. The next two terms are biological. The last two terms are, in a sense, sociological and bear on the capabilities and circumstances of a civilization.

When the equation was first discussed, guesses to the values of all the terms in the equation yielded a huge range of possibilities. One of the challenges of the equation is that there are a lot of factors going into it, and each is subject to a large uncertainty. In his autobiography Drake recalls closing the meeting: "We've reached a conclusion, there are somewhere between one thousand and one hundred million advanced extraterrestrial civilizations in the Milk Way."[9]

There are radio astronomers who regularly spend a fraction of their time looking for SETI signals, and some have committed their lives to it. There are also now multiple dedicated SETI searches, often financed by private individuals and corporations. Signal candidates do emerge, but the sources are usually found to be human-induced or a rare, unexplained, one-time occurrence.

One of the most famous discoveries was a strange periodic radio signal found by graduate student Jocelyn Bell in 1967. The team she worked with had no idea what it was initially, and a signal from an extraterrestrial civilization was as good an explanation as any. The signal flickered every 1.3 seconds and was unlike anything seen before in radio astronomy. Bell and coworkers called the signal LGM-1, for "Little Green Men 1." It turned out to be a significant astronomical discovery of a phenomenon known as a *pulsar*, which is the remnant of a star that spins rapidly with a directional beam of radiation that gets emitted, giving the regular signal characteristic.

Another famous episode is the Wow! Signal. Ohio State University operated a large radio telescope near Delaware, Ohio. In one of the longest-running dedicated ET searches, researchers scanned the sky near the 21cm hydrogen line. In 1977 astronomer Jerry Ehman, who was working on the OSU SETI project, was pouring over printouts from a scan, and saw a huge signal, and wrote "Wow!" next to it. There is still no known explanation for that signal, but there were some problems where only one of two radio signal feeds picked it up. Attempts to find the signal since have come up empty, and it is more likely that it was just a statistical fluke, particularly given that it was only on one radio feed.

The Wow! Signal made it into a few instances of popular culture. In the *X-Files* episode "Little Green Men," a mention was made of it. The strength of the signal in the printout came out as 6EQUJ5, which is an encoding of the signal strength. The numbers represent a weaker signal, and then diving into the alphabet gets a stronger signal, with U being the largest. 6EQUJ5 appears in a number of cultural appearances, like the movie *Ad Astra* where it's the filename of a top-secret message.

A more recent SETI candidate was detected by the Parkes Radio Telescope in Australia in April and May 2019 while scanning Proxima Centauri, the closest star to Earth. Proxima Centauri is a red dwarf with two planets, one of which is close to the Earth's mass. The radio signal showed up over several days in five thirty-minute intervals and had a directionality to Proxima. When pointed away from Proxima, it went away. The detection was part of a SETI search called Breakthrough Listen and was tagged as BLC-1 for Breakthrough Listen Candidate 1. A subsequent analysis of BLC-1 suggested that it has a human origin.[10]

SETI has gained considerable interest from the public with private sponsors. Paul Allen, the cofounder of Microsoft, helped fund the construction of the Allen Telescope Array (ATA) located in Northern California. One of the major motivations for the ATA is SETI. The telescope design minimizes false positive signals from terrestrial radio sources.

Radio frequencies aren't the only path for SETI. One possible direction is looking for laser signals beamed from outer space. Normal light from stars has a spectrum of frequencies that reach us, but lasers produce monochromatic light. Lasers can transmit information much faster than radio waves, so it may be a considerable advantage to broadcast using visible light. Detection of laser sources can be done with relatively inexpensive cameras. The laserSETI project is a crowd-sourced effort to deploy inexpensive cameras across the globe to provide full coverage of the night sky.[11]

The flip side of SETI searches are broadcasts into outer space, where we transmit rather than listen. Since the development of radio and subsequently television, we've been broadcasting into the void for a century. Most radio and television transmissions on Earth spread out in space from the transmission towers and are probably just too weak to reach the ears of a listening alien. However, as shown in figure 6.2, if you invert the radio telescope to become a transmitter, you can produce a narrow beam of radio waves to send toward a distant target. While the "listening" part is sometimes referred to as passive SETI, active SETI is broadcasting.

An early broadcast example of active SETI was done at the Arecibo Radio Telescope in Puerto Rico in 1974 by Frank Drake. It targeted the Hercules star cluster and could be strong enough to be received by a listening alien civilization if they are tuning in. The cluster is 24,000 light-years away, so it will be quite some time before it reaches the

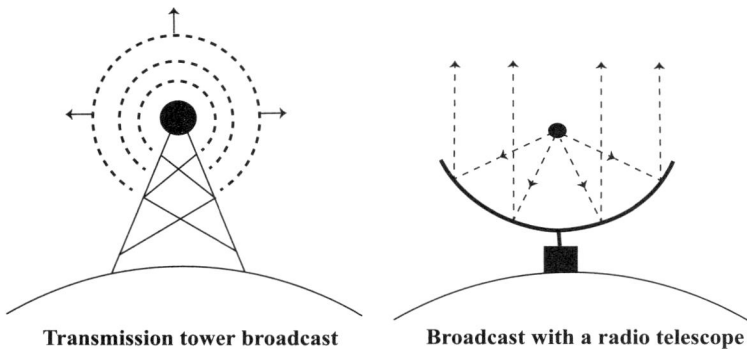

Transmission tower broadcast   **Broadcast with a radio telescope**

FIGURE 6.2 Radio waves spread out from a transmission tower, but broadcasting from a radio telescope concentrates the outgoing radio waves into a directable beam.

target. This first broadcast was part of a reopening ceremony for the Arecibo Telescope after a substantial reconstruction project.

The concept of active SETI has come under significant criticism because it might give away our presence to a hostile civilization, giving rise to a real "War of the Worlds" invasion. In his novel *The Three-Body Problem*, author Liu Cixin writes about a scientist who projects a message from Earth into space using the Sun as a transmission amplifier. The Earth then falls under scrutiny of an advanced civilization in the nearby Alpha Centauri star system. In his sequel, *The Dark Forest*, he warns of the dangers of active SETI:

> The universe is a dark forest. Every civilization is an armed hunter stalking through the trees like a ghost, gently pushing aside branches that block the path and trying to tread without sound. Even breathing is done with care. The hunter has to be careful, because everywhere in the forest are stealthy hunters like him. If he finds other life—another hunter, an angel or a demon, a delicate infant or a tottering old man, a fairy or a demigod—there's only one thing he can do: open fire and eliminate them. In this forest, hell is other people. An eternal threat that any life that exposes its own existence will be swiftly wiped out. This is the picture of cosmic civilization. It's the explanation for the Fermi Paradox.

In his prosaic way Liu reveals the solution to the Fermi paradox: no one wants to reveal their presence and become a target. Lest we imagine that civilizations out there might be listening and friendly, we can look to historical examples on Earth. When civilizations collided, it often got ugly.

As with radio waves, lasers could conceivably be used in active SETI. Investigators at MIT have proposed that a huge laser could be retasked to project its energy through a telescope, and that the resulting beam would be detectable by a civilization as far as away as the TRAPPIST-1 star, some forty light-years away.[12] TRAPPIST-1 has seven exo-planets, including three in a potentially habitable zone.

THE PIONEER PLAQUE

If we were to somehow establish contact with an alien civilization, the natural question arises: how do we communicate? As we saw above, Cocconi, Morrison, and Drake speculated that the 21cm wavelength of hydrogen might be a natural frequency that other cosmic civilizations might naturally use as a broadcast channel. Falling back on fundamental constants of nature and known galactic structures could provide a cosmic Rosetta Stone to form a basis for communications. Carl Sagan, with Drake's help, put this concept into action with their design of a plaque that was placed on the Pioneer 10 and 11 space probes.

Pioneer 10 and 11 were built as probes of the outer planets and utilized a gravitational slingshot effect to propel them beyond the solar system. The plaque Sagan and Drake designed is shown in figure 6.3. In the upper left-hand corner is a diagram that is supposed to describe the physics behind the hydrogen 21cm transition, and also

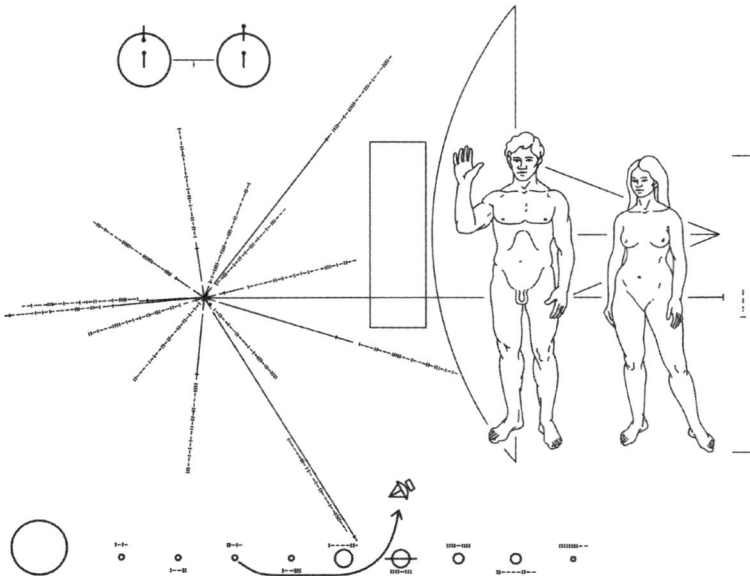

FIGURE 6.3 Plaque on Pioneer 10 and 11.

set a distance scale using 21 centimeters as a basic unit of distance and 0.7 nanoseconds as a time measure.

The markings on the plaque are in binary, which assumes that 1s and 0s are a basic universal number system. The standing man and woman are shown next to a drawing of the spacecraft itself, which sets a scale for our statures. The bottom of the plaque displays the solar system, and the path of Pioneer, including its origin on Earth. The binary numbers below the planets indicate the relative distance to the Sun.

Perhaps the most interesting piece of the plaque is the diagram with fifteen radiating spokes on the middle left. The longest one is intended to indicate the distance from the Sun to the center of the galaxy. The remaining spokes represent pulsars, with their periods and relative directions and distances indicated in binary. This uses the pulsars as a cosmic GPS system to locate the solar system.

The odds are exceedingly slim that either of the plaques will be found by an intelligent civilization. Even though they have exited the solar system, their speeds and trajectories don't put them in any likely position to be discovered. The last communication from Pioneer 10 was in 2003.

INTERSTELLAR PROBES

Beyond SETI, listening in on radio frequencies, or looking in the optical band, is the question of physical probes that travel from one star system to another. But the challenge with probes is the vast distance scale of space.

Compared to the scale of the solar system the expanse of space, even to the nearest stars, is difficult to ponder. To give a feel for this, Proxima Centauri is the closest star to our solar system and is 4.2 light-years away, meaning that a pulse of light from Proxima takes 4.2 years to reach us. The Sun is 93 million miles from Earth. Proxima, our *nearest* neighbor, is 25.3 trillion miles away.

Normally, spacecraft can only reach a tiny fraction of the speed of light. At the time of this writing, the fastest velocity achieved by a manmade object in space is the NASA Solar Parker Probe, that attained 100 kilometers per second (60 mi/s) when it crashed into the Sun. But, even at that speed, it would take over 6,000 years to reach the nearest star.

For science fiction, the distance between stars, or galaxies for that matter, has been a challenge for plot development. Various tricks have been put in place to circumvent this limitation, like warp drive in *Star Trek*, or the use of wormholes in the movies *Contact* and *Interstellar*.

Soon after Cocconi and Morrison's paper on SETI with radio telescopes, electrical engineer Ronald Bracewell proposed interstellar probes as an alternative to two-way radio communication in a 1960 paper.[13] In his model an advanced civilization would launch multiple probes to stars that were likely to harbor intelligent life. The probes would have to be highly autonomous, possibly possessing artificial intelligence, and would have to be patient, waiting for a civilization to develop significant technologies. A probe could be sent to a distant star or planet and park there. This "parking" would greatly reduce the time and power needed for communication between the developed civilization and the probe. In one scenario the Bracewell probe would wait to receive a radio transmission from a planet it was monitoring, and then would echo back the signal to the planet, perhaps also relaying to "home" that it had found intelligent life. One criticism of the idea of a Bracewell probe is that it is highly resource-intensive.[14]

The movie *2001 A Space Odyssey* used a Bracewell probe as a plot device. Buried in the Moon, a mysterious monolith is excavated by humans and starts to beam signals back home. The idea being that if humans found this device, they had developed the capabilities of space flight, activating the probe.

A proposal for an Earth-initiated interstellar probe comes from the Starshot Breakthrough Initiative, which was formed to create the proof-of-concept of a light-sail propelling a tiny spacecraft on a voyage to the nearby star Alpha Centauri. Cofounded by philanthropist Yuri Milner, physicist Stephen Hawking, and Meta head Mark Zuckerberg in 2016, the goal is to send ultralight cameras to photograph Proxima b, a planet found in the Centauri star system. This would be on a timescale of twenty years after launch, not thousands of years. At its highest speed, it would be traveling at roughly 20 percent of the speed of light. Upon reaching Proxima b, the spacecraft would beam images back to Earth.

Returning to Enrico Fermi's question, "Where are they?," there are the myriad reports of unidentified flying objects (UFOs) attributed to aliens. Believers in extraterrestrial visitations and government cover-ups vastly outnumber flat-Earthers. While most scientists consider such reports fringe, the possibility of a systematic study of UFOs has gained significant political currency in the US Congress and in government agencies like NASA.

The abbreviation "UFO" has been replaced with "UAP," an abbreviation for Unidentified Anomalous Phenomena. A significant fraction of the UAPs are sightings of what appear to be objects undergoing anomalous accelerations or traveling at unexplained high velocities. NASA commissioned an independent study panel to report on UAPs in 2022, with a report issued in 2023. They concluded:

> To date, in peer reviewed literature, there is no conclusive evidence suggesting an extraterrestrial origin for UAP, the challenge we have is that the data needed to explain these anomalous sightings often do not exist; this includes eyewitness reports which, on their own can be interesting and captivating, but aren't reproducible and usually lack the information needed to make any definitive conclusions about a phenomenon's provenance.[15]

The report recommends that NASA provide a coherent, systematic framework to collect and analyze UAP reports. The systematic analysis is not limited to UAPs on Earth but can be extended to space. Two objects of interstellar origin have been sighted passing through the solar system. One is a curious object called 'Oumuamua, observed in 2017. The second is a lesser-known Comet Borisov that passed through in 2019. Although there was a suggestion that 'Oumuamua was a deliberate interstellar probe, there are natural processes that can produce these extrasolar interlopers. A new facility coming online, the Vera C. Rubin Observatory (VRO), should be capable of detecting more of these in the near future. The VRO has a very high data throughput and can scan the entire visible sky over the course of a number of nights and would be sensitive to more transient interstellar objects.

While extraterrestrial intelligence seems plausible, there are still no real smoking guns since the time of Galileo, Kepler, and Huygens, when we first trained telescopes on the planets. However, an essential legacy of these early pioneers was a seismic shift in our modern understanding of space.

* 7 *

ON EARTH *as* IT IS
in HEAVEN

Until now, we have dealt with spatial perspectives where the objects occupying the space defined it. Moreover, the conceptions of space are largely geometrical. We had Aristotle's model of nested spheres with the Earth at the center, or Ptolemy's variation. The spaces of Copernicus and Aristarchus were defined by the Sun at the center. The possibility of intelligent extraterrestrials emerged from the observation of structure on planets enabled by the telescope.

But, with the advent of Newton's physics, we have a space and time not defined by the objects that inhabit the space, but as a space with intrinsic properties that are defined by physical interactions. This is akin to having a stage where props and actors have yet to be defined, but the stage itself has its own rules. In this regard, the concept of space that emerged from Newton was yet another leap toward a kind of extreme allocentrism, where there is not an up or a down or a north or a south. There is no center or periphery.

The emergent laws of motion are called classical mechanics, which was spawned from the realization that the nature of motion on Earth was the same as motions in the heavens. From the perspective of someone in the seventeenth century, this in and of itself is revolutionary. Also, the emergent vision of space was driven from empirical observations, which makes it even more innovative. While Aristotle was of the opinion that one could make progress with pure reason, the new vision of space and the laws of motion were empirically derived. In other words, we had to listen to Mother Nature to discern the structure of space and time, and not project our beliefs.

The subsequent developments of space in the twentieth century are relativity and quantum mechanics. We'll get to this next, but there is frequently a misconception that these more recent developments superseded classical mechanics. That's not the case. Classical mechanics is viewed as a limiting case that relativity and quantum mechanics must reduce to under certain conditions.

There are two important paths to this synthesis. First, there is motion on Earth, for which Galileo gets much credit. Second, there is motion in the sky, for which Kepler gets much credit. Finally, there is the synthesis of Newton that both motions are the same and universal laws of motion can be written down in a systematic manner. A model of space is the key ingredient.

THE A-WORD

One of the fundamental barriers to understanding how forces work was the realization that they produce acceleration. For students taking physics for the first time, the concept of acceleration can sometimes be challenging to comprehend. Likewise, the challenge was something of a historical barrier.

Aristotle held that forces produce velocities. In everyday life this seems natural. If you're pushing a wheelbarrow, it rapidly grinds to a halt as soon as you stop applying force to it. If you push the wheelbarrow harder, it moves at a faster velocity.

Velocity is a combination of a directionality and a speed. To specify a velocity, you must call out both. "Traveling at 65 miles per hour" gives a speed. "Traveling northeast at 65 miles per hour" gives both a direction and a speed, so this description gives a velocity.

More grandly, we can think of a velocity space. Physicists rarely use the term "velocity space," but it is a concept we can entertain. You can think of an object at rest with respect to the Earth as having zero velocity but, depending on the speed and direction, an object may occupy some spot in this velocity space. A car traveling northeast at 65 miles per hour is occupying a single position in that velocity space, provided its velocity doesn't change: its direction and speed are constant. Although it may be traveling through what we think of as a physical space, it is stationary in this velocity space.

Now, how would an object move from one position in velocity space to another? It would have to change its velocity: to change either its speed or its direction of motion or both. How an object moves around in velocity space is acceleration.

Let's focus on units. Velocity is how an object moves in our physical space from one location to another over some length of time, hence velocity is distance traveled/time it takes, where the units could be miles per hour or meters per second. Likewise, an object moving from one spot in velocity space to another traverses different velocities over some time interval, and this "velocity of velocities" (acceleration) is the change in velocity divided by the time it takes. The units might be miles per hour per second. If a car is stopped at a stoplight, and then suddenly the light changes and the driver floors the accelerator, we might say that he goes "from zero to sixty in ten seconds." In effect, the car is going from one spot in velocity space (zero) to another spot (60 miles per hour) in ten seconds. Since physicists like units of meters, the standard units of acceleration are (meters/second)/second or, more compactly, m/sec^2 (spoken as "meters per second-squared").

In Aristotle's spaces, the "zero" or origin of physical space was the center of the Earth. Likewise, with Aristotle claiming that forces are necessary to create velocity, the natural origin or zero of velocity space would be zero. In Newton's version of space, there is no preferred "zero" or origin in physical space, nor is there a preferred "zero" or origin in velocity space, but there is a preferred "zero" or origin in acceleration space, namely zero acceleration.

GALILEO AND ACCELERATION

A general concept of Aristotle was that "truths" could be discovered through reason. Around the time of Galileo, there were cracks in this bedrock with experiments and observations leading to new truths. Although it's difficult to define "the scientific method" in an unambiguous fashion, the emerging use of observation and measurement coupled with numerical analysis was one of Galileo's signatures.

Galileo's attitude toward experiment and science is famously articulated in one of his letters, in something of a swipe at the authority of Aristotle in scientific matters:

> For in the sciences the authority of thousands of opinions is not worth as much as one tiny spark of reason in an individual man. Besides, the modern observations deprive all former writers of any authority, since if they had seen what we see, they would have judged as we judge.[1]

Galileo found that objects did not fall with a constant velocity, but rather with a constant *acceleration*. Moreover, the acceleration was independent of the object's mass.

There is the famous story of Galileo dropping balls with different weights from the leaning Tower of Pisa to demonstrate that two objects fell with the same rates, regardless of weight. Historically, this may not have happened, but it is an illustrative demonstration.

The significant experiment that Galileo did perform was with a ball rolling down a gently inclined slope, which allowed him to time the rate of descent with the instruments available in his era. He found that the distances the ball traveled went like the square of the time in descent. So, if the ball descended a certain distance in one unit of time, it would descend one quarter of the distance in half the time. He performed this experiment over varying distances, times, and inclinations, and the relation of the distance of descent and time always went as the square of the time.

Why does the distance traversed going as the square of time imply acceleration? If the pull of gravity produces a constant velocity, the time it takes to traverse different distances would be proportional to time. If an object traverses one cubit (the unit Galileo used) in one second, it would traverse two cubits in two seconds if the velocity was constant. On the other hand, if it traverses four cubits in two seconds, then the velocity must be increasing over time. Hence, it's accelerating.

Galileo published his findings in his book *Discourses and Mathematical Demonstrations Relating to Two New Sciences*. Here he argued that the objects falling under the influence of Earth's gravity experience a constant acceleration, regardless of mass/weight.

One consequence of this motion due to gravity is the path of a projectile near the surface of the Earth, called a parabola. The horizontal motion of the object, say a cannonball, may be a constant speed, while the vertical motion has a changing speed that scales as Galileo found in his experiments.

This study of motion of objects on Earth was more than just an academic interest, it had and has substantial utilitarian benefits. By calculating the path of a cannonball, an army officer could direct the projectile onto an enemy target. A representative parabolic path for the cannonball is shown in figure 7.1. The angle of the barrel of the

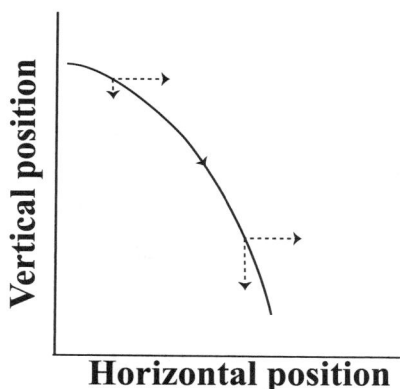

FIGURE 7.1 Parabolic path taken by a cannonball near the surface of the Earth. The dotted arrows represent the horizontal and vertical velocities. The horizontal velocity remains constant, but the vertical velocity changes due to acceleration.

cannon and the speed of the cannonball as it leaves the cannon determines where it falls. This understanding had an impact on the use of artillery: how to aim a projectile to hit a target. Although the intent of scientists is usually coupled with pure knowledge, time and again we find that new discoveries are often co-opted for military advantage.

FRAMES OF REFERENCE

A "frame of reference" is the concept of a set of geometric axes where the positions of objects can be measured at various times. Think of it as a kind of three-dimensional scaffolding onto which you can hang observations, like the unfolding of the position of a flying cannonball over time. There can be multiple frames of reference where multiple observers can witness the unfolding of events in these different frames.

We can ponder what it means to make observations in different frames of reference. Imagine you are seated on a train in a station, where the train you're sitting on is one frame of reference and the platform is a second frame of reference. Perhaps there's another train across the platform that effectively becomes a third frame of reference. Events, like a person putting luggage in an overhead compartment, can be referenced to your train, the platform, or the second train. You could see the person put luggage in the overhead from the seat next to them, or through the window of the train from the platform, or even from the train across the platform.

Every so often, when I'm in a parked train in a station and I see a second train next to me move, I get momentarily confused as to whether I'm moving or the other train is moving, and I need to look at something in the station to figure out who's moving. Each train and the platform constitute a frame of reference.

Now both trains might be moving through the station at different speeds. Let's imagine that they aren't accelerating at all, but just moving with a constant velocity. Although we'd be tempted to say that the platform is not moving, we can just say that its speed is zero, although it might be moving backward from the point of view of a person on a train. All three—the two trains and the platform—all form frames of reference. The one thing that all three share is a common time. That is to say, the clock on one train, the clock on the platform, and the clock on the other train tell the same time. This changes when we consider Einstein's relativity, but let's leave that alone for now.

In the Aristotelean notion of space, we might think that the station platform was somehow privileged because it represents the Earth. But when nature only recognizes acceleration as due to forces, there is no privileged frame of reference.

You can imagine measuring things from the vantage point of either train or the platform. Say someone is running down the platform and trips and falls. The passengers on the trains and a witness on the platform would all agree that the runner falls at the same time. Since the runner encounters a force that makes him trip, he falls under the influence of gravity and we see him accelerate downward in the fall. The people on the train and on the platform would all see the same change in velocity of the poor runner and would figure out that he was subject to the same force as seen from all vantage points. The speeds of the train don't make any difference, they all see the same force and consequent downward acceleration in the fall.

A central concept to the emerging new physics of the Galileo/Newton era is that the laws of physics are identical in all reference frames moving with respect to each other with a constant velocity. This is sometimes called Galilean relativity, to distinguish it from Einstein's relativity, which will come up next. Comparing events from the perspective of one frame of reference to another is often called a transformation.

The frame of reference described here is also sometimes called an "inertial frame of reference" because they move with constant velocities relative to each other. What else could there be? In an accelerating car you might feel yourself pushed back into your seat, as if a force is pushing on you. You aren't really being pushed, it's your body's resistance to acceleration. In this situation, your accelerating car is a non-inertial frame of reference.

In the first chapter I introduced the binary concept of egocentric and allocentric frames of reference. It's perhaps natural for the reader to wonder whether this new vision of space fits into either category. Recall that the egocentric/allocentric divide is associated with landmarks and navigation. In a space that lacks landmarks or objects, it is a bit pointless to ask whether it's intrinsically one or the other. But we can imagine populating the space with different objects and see where this notion leads. In the case of two trains and a platform, all motion is relative and there is no privileged frame of reference for any observer, so it's tempting to label this space as allocentric. Moving forward, we can use this generalization as a principle for analyzing concepts of space in this framework of imagined observers.

ABSOLUTE TIME AND SPACE

In the transformation rule above, there is a phenomenon we're accustomed to: time is the same in all frames of reference. In that sense Newton postulated a concept that he called "absolute time"— meaning that there is a kind of universal clock that ticks at a uniform rate for everyone, everywhere.

Newton also spoke of a concept of an "absolute" space, which would seem to imply that there is a preferred frame of reference in the universe that defines things at rest, like we might consider the train platform in the above example. Although he articulated this phrase of "absolute space," his formulation of the laws of motion do not actually create a fixed or privileged frame of reference, while the absolute time *is* embedded in his theory. There *is* a kind of "absolute" in the sense of acceleration, however. That is to say, the universe seems to put some importance to a "zero" of acceleration.

MOTION IN THE HEAVENS

Another hangover from Aristotle was the entire system of motion in the skies. Ptolemy demonstrated a resilient method of calculating the positions of the Sun and planets with an Earth-centered solar system and perfect circles, albeit with displaced centers and epicycles. As mentioned above, this was used to create almanacs that were mostly used for astrology, but these gradually came into more and more use with celestial navigation. Regardless of the usage, the accuracy of the almanacs was an important driver of change.

Copernicus published *On the Revolutions of the Celestial Spheres* in 1543. This had the famous model of the planets orbiting the Sun. The first hints of his Sun-centered model of the solar system are dated to around 1514. *On the Revolutions* was effectively completed in 1532, but he strongly resisted publishing up until just before his death. The resulting tables of planetary positions were roughly as accurate as the Ptolemaic system, although there were differences in the details.

Danish nobleman Tycho Brahe developed a large and precise astronomical observatory on the island of Hven. Brahe was highly interested in figuring out whether the Ptolemaic or the Copernican system was more accurate. For what follows on Brahe and Kepler, I owe much gratitude to my late colleague Owen Gingerich, who investigated the journals and writings of both.[2]

Brahe realized that, at the point of closest approach, the distance from Earth to Mars would be half the distance predicted by the Ptolemaic system in the Copernican model (see fig. 7.2). For reference, an astronomical unit (AU) is the Earth–Sun distance. We now know that this is about 150 billion meters, but the distance scale was not known at this time. Brahe set out to measure the Earth–Mars opposition distance in a technique called diurnal parallax, which is looking at the position of Mars just after sunset and just before sunrise. By using the Earth's diameter as a baseline, he could, in principle, see a difference in the position of Mars that would rule out one or the other system and set a scale for the solar system. Unfortunately, Brahe's instruments weren't sensitive enough to observe the small shift, and refractive effects in the atmosphere also spoiled it. Brahe dropped this idea.

In the 1593 opposition Brahe observed a significant deviation in the position of Mars from both the Ptolemaic and Copernican predictions. Specifically, the Ptolemaic model was out of the observed longitude by +5 degrees, and the Copernican model was out by

−4 degrees. Brahe's precision was about 2 arc-minutes (1/30th of a degree), so this discrepancy was highly significant. Every thirty-two years, the position of Mars was out of kilter by this amount in both the Copernican and Ptolemaic systems.

In 1600 Brahe moved to Prague and hired Johannes Kepler as an assistant. Kepler was a math teacher at a Lutheran high school but found himself out of work. Brahe gave Kepler the problem of understanding the position of Mars. In the Copernican model the planets had eccentric orbits and sped up when they were closer to the Sun, but the one important exception to this construction was the Earth, which Copernicus assumed moved at a uniform speed. Kepler

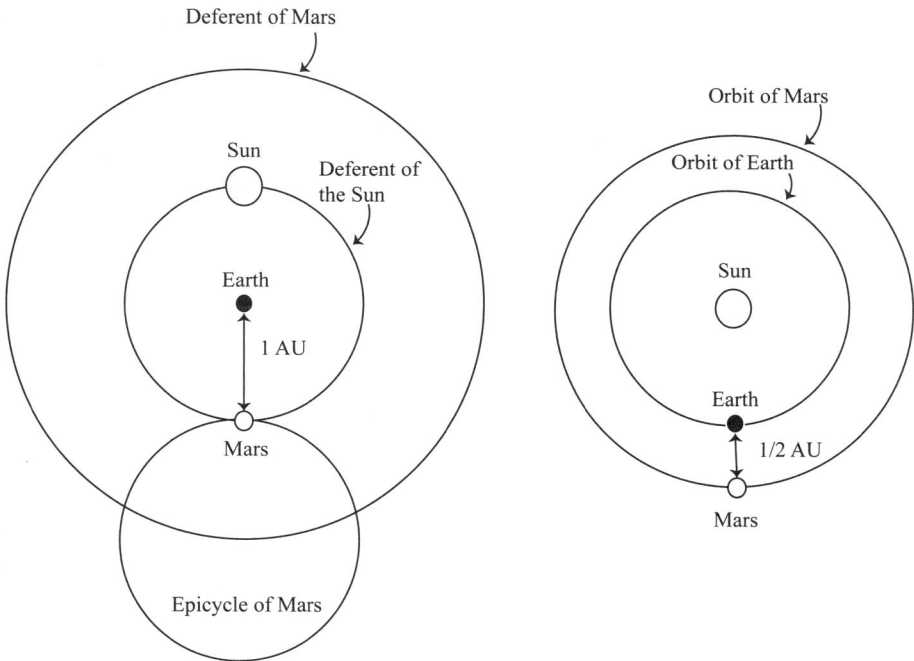

FIGURE 7.2 The left diagram shows the orbit of Mars in the Ptolemaic system; the right shows the orbit of Mars in the Copernican system. At the point of closest approach to Earth, Mars is 1 AU (astronomical unit—equivalent to the average Earth–Sun distance) away in the Ptolemaic system, whereas in the Copernican system it was half this distance.

questioned why the Earth would be an exception to the motions of the other planets. An important point is that the position of Mars in the sky depends both on Mars in its orbit and where the Earth is when observing it. Kepler released the constraint of the Earth moving at a uniform speed and experimented with different oval shapes to get a more precise formulation.

Kepler had Brahe's observations to try to fit into his new formulation. After much effort, Kepler hit on a model where the planetary orbits were ellipses that made a great fit to the position of Mars in the sky over time, particularly at the oppositions.

A note is in order of the construction of ellipses. One property of an ellipse is the presence of two foci. If I were to ask you to draw a circle as best you could, you might take a string and pin it in the center of the circle. You then attach a pencil to the end of the string and, keeping the string taut, draw out the circumference of the circle. Ellipses have, in some sense, two rather than one center, called the foci. You can trace out an ellipse by taking a loop of string and pin it at each of the foci. The pencil holds the string-loop taut and will trace out the curve of the ellipse (fig. 7.3).

Kepler decided that the offset of the orbits could be accomplished by placing the Sun at one focus of the ellipses representing Earth's and Mars's orbits. He additionally created rules for how the planets slowed down or sped up in their orbits: they moved fastest at the point of closest approach to the Sun and slowest at the farthest distance.

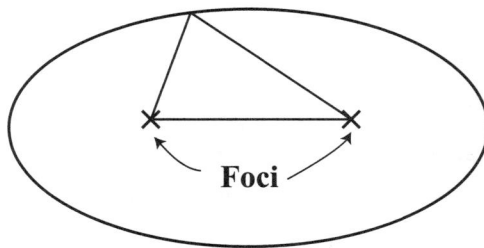

FIGURE 7.3 Construction of an ellipse. If you take a loop of string and fix it to the two foci of the ellipse, the taut end of the string will trace out the path of the ellipse.

Finally, there was an overall constant that governed the speed of the orbit depending on the distance from the Sun.

With these effects considered, the tables Kepler produced gave predictions for the position of Mars to an accuracy consistent with Brahe's precision for measurements. His construction also solved the once-in-thirty-two-year catastrophe of Mars being out of position by five degrees.

Kepler's choices were informed by his belief in physical causes of the motions of the planets. He used the word "focus" to describe the location of the Sun in the elliptical orbits, as *focus* means "hearth" in Latin. So Kepler placed the Sun as the "hearth" of planetary orbits. His book on planetary motions, *Astronomia nova* (new astronomy) was subtitled *Aitiologitos, seu physica coelestis*, "based on causes, or the celestial physics."[3]

In *Astronomia nova*, Kepler writes:

> I am much occupied with the investigation of the physical causes. My aim in this is to show that the celestial machine is to be likened not to a divine organism but rather to a clockwork . . . , insofar as nearly all the manifold movements are carried out by means of a single, quite simple magnetic force, as in the case of a clockwork all motions [are caused by] a simple weight, moreover I show how this physical conception is to be presented through calculation and geometry.[4]

Here we have the hints at what we call *action at a distance*, namely forces that are transmitted through space.

As a coda, in 1672 astronomer Giovanni Cassini was able to measure the Earth–Mars distance using parallax. By this time the modern model of the solar system was widely accepted. By measuring the Earth–Mars distance, a physical value for the astronomical unit (AU) could be determined. This and subsequent measurements revealed just how empty the solar system is. The Sun's diameter is about 1/100th the Earth–Sun distance for example. And, yet somehow the force of gravity is transmitted through this vast empty space.

CELESTIAL MOTION

While Galileo showed the importance of acceleration arising from forces, Kepler showed how an approach based on presumed physical causes could generate highly accurate predictions of celestial motions from a small set of principles. The question of action at a distance naturally arose as there was no physical contact between the Sun and the planets, unlike the idea that forces are transmitted from the Prime Mover inward through the planetary spheres.

But what was the nature of this force? A major clue was in the speed of the planets in their orbits. Saturn is ten times further from the Sun than Earth and it has an orbital period about thirty times ours. The speculation was that the attractive force to the Sun got weaker and weaker with distance. One French scientist, Ismael Boulliau, speculated that the force from the Sun fell off like the inverse square of the distance and suggested that this would give rise to Kepler's results. This would imply that the gravitational force from the Sun on Saturn is 1/100th the force on Earth.

The speculation that gravity followed an inverse-square law became widespread. In England astronomer Edmund Halley shared this notion, but no one could prove that the inverse-square law would lead to Kepler's elliptical orbits. Halley traveled to Cambridge and approached Isaac Newton to see if he could prove that this was the case. Newton claimed to have already demonstrated it but couldn't find his papers. Newton then promised Halley that he would send a proof back to London when he finished the calculations.

Newton used Galileo's acceleration as a cornerstone of how forces work and developed the calculational tools necessary to deduce motion from forces in general, and then applied the tools to the inverse-square law. He had to first articulate his famous three laws that are the foundation of his description of forces and motion. Two of the three laws were grounded in what I discussed above: acceleration and frames of reference. Even though most folks know these laws and they may seem utterly familiar, I'll write them down anyway:

LAW 1: A body in motion stays in motion, a body at rest stays at rest.

This is mainly a statement that a uniform velocity is a kind of frame of reference—objects travel in a straight line with constant speed if no force is present.

LAW 2: Force equals mass times acceleration.

This acknowledges what Galileo found about acceleration being fundamental to forces.

LAW 3: For every action, there is an equal and opposite reaction.

The third law is a statement of symmetry, that there is no privileged "object"—a kind of democracy among objects. In essence, if you push on something, it pushes back by the same amount.

But these laws were insufficient to link the inverse-square law to the planetary orbits. To get to his proof, Newton had to develop a mathematics describing continuous motion: calculus.

Ultimately Newton published the proof in his famous work, *Philosophiae Naturalis Principia Mathematica* (Mathematical Principles of Natural Philosophy), better known as *Principia*. There was much more in *Principia* than just the proof of the elliptical orbits.

Newton also showed that the attraction of gravity at the surface of the Earth was constant, as we're at a uniform distance from the center. This gives rise to the parabolic trajectories seen in cannonballs. Going beyond this, Newton made a thought experiment of a cannonball fired from a tall mountain on Earth (fig. 7.4). When fired with a relatively small initial velocity, it falls in the familiar parabolic path. But, as the initial velocity increases, the cannonball falls, but misses some of the Earth because of its curvature. With a sufficient initial velocity, the cannonball is continually falling, but also continually missing the Earth, causing it to orbit around, in the same way that

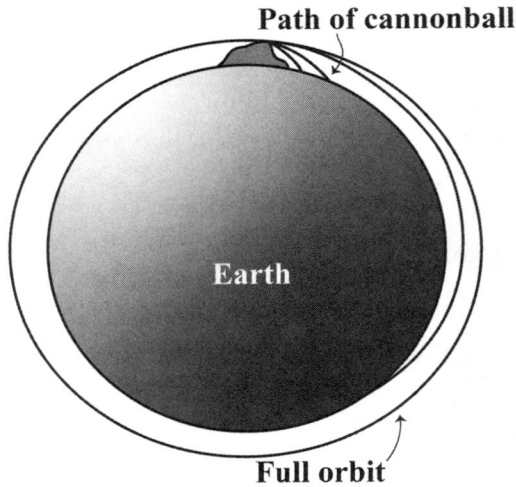

FIGURE 7.4 Newton's thought experiment about a cannonball fired from atop a mountain on Earth, where both a parabolic trajectory and an orbit can be seen, depending on the initial velocity of the cannonball.

the Moon orbits the Earth or the planets around the Sun. The motion on Earth and in the heavens was unified. We inhabit the same space with the same rules of motion.

ACTION AT A DISTANCE

One important wrinkle in Newton's gravity is the concept of *action at a distance*. While we are accustomed to objects moving when pushed on and there is physical contact, there is no physical contact between the Sun and a planet, yet the planet will feel the gravitational force of the Sun. This force is somehow transmitted through empty space between two objects. How to explain or even imagine this?

Over time we've come to visualize the action-at-a-distance forces as having field lines that give a map of how they spread out in space. Although the field lines are invisible to the eye, we can imagine placing a "test" mass somewhere in the force field, and the direction and intensity of the field lines give a prescription for where the force is pushing the test mass and how hard (fig. 7.5). By probing space

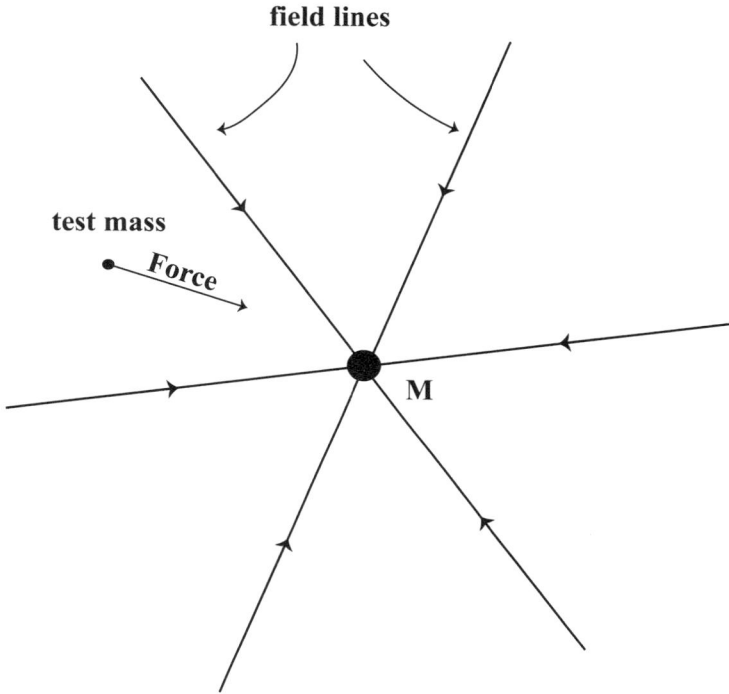

FIGURE 7.5 Field lines spreading out from a mass, *M*, and the placing of a test mass to find the strength and direction of a force. The force of gravity is attractive, so the field-line arrows are drawn pointing at the mass in the center.

with test objects we can develop a true map of where the field lines are going.

The field lines spreading out from a point-like object are the origin of the inverse-square law that Newton demonstrated led to Kepler's elliptical orbits. As the field lines spread out all over space, they become less and less dense, resulting in progressively less force as you move further away from an object.

Although this is not a physics text, I'd be remiss not to mention electricity and magnetism. Electricity and magnetism can also be described by forces that are drawn as action-at-a-distance field lines. While the source of field lines in gravity is the mass of an object, the source of the field lines for the electric force is charge. While there is only one kind of mass (as far as we know), there are two electrical charges, but there are significant parallels to gravity. For one, the electric field lines for point-like charges spread out like the inverse-square law for gravity. Also, the electrical force is directly related to the amount of charge, as the force of gravity is directly related to the mass of an object.

In the nineteenth century the electric and magnetic forces were shown to both originate from charges. While a fixed charge creates an electric field, a moving charge also produces a magnetic field. Ultimately these were shown to be manifestations of the same electromagnetic force, although we often draw electric and magnetic field lines separately.

While electromagnetism was developed quite some time after the publication of *Principia*, the fundamental principles of motion articulated by Newton persisted, and the idea that space was filled with lines of force communicated by objects became the norm. But the effect on the human psyche was considerable. Kepler already conceived of the universe as a kind of clock with fixed laws determining the fate of all.

Even on the scale of the solar system space is quite empty, yet it conveys the force of gravity through the void. Both the newly realized scale of space and the mechanistic nature of the clockwork universe are daunting, as it already seems foreign to our experience of human intentionality. Author Gale Christianson, who wrote on the development of Newton's theory of gravity, put it this way:

> When we step back and look at Newton's universe from afar, what is it, exactly, that we see? According to the *Principia*, we peer into a seemingly endless void of which only a tiny part is occupied by material bodies moving through the boundless and bottomless abyss. Newton's followers would liken it to a colossal machine, much like the clocks located on the faces of medieval buildings. All motions are reduced to mechanical laws, a universe where human beings and their world of the senses have no effect. Yet for all its lack of feeling, it is a realm of precise, harmonious, and rational principles. Mathematical laws

bind each particle of matter to every other particle, barring the gate to disorder and chaos.[5]

PEERING INTO THE VOID

In the *Divine Comedy*, a journey through the planets to the firmament seemed like a stroll in the park for Dante, but the history of space is an expedition into a progressively larger and larger void. Arguments in favor of a Copernican model were that the stars are so distant that they have no apparent motion as the Earth orbits the Sun. From the seventeenth through the present, we've seen just how vast space really is.

I mentioned above that Cassini was able to measure the Earth–Mars distance in 1672 using a technique called parallax. The concept is relatively straightforward. If you're driving down a road and see a distant hill, it doesn't seem to move much, but in the foreground telephone poles and houses move by more rapidly. The difference in angles that you see to objects depends on their relative distances and the vantage points from which you view them. You can also try this by holding out a finger at the end of your outstretched arm and looking at an object with one eye or the other and see an apparent change in position.

Cassini used two locations on the Earth to establish a baseline for his parallax measurement of the Earth–Mars distance. Over the next century and a half telescopes improved dramatically, and the orbits of the planets and distance scales within the solar system were established. Knowing the scale of Earth's orbit astronomers began to use this as their baseline for parallax (fig. 7.6). By the 1830s astronomers could start to measure the distances to close stars by measuring how much their positions changed from different viewing points in our orbit.

A parsec is the distance at which the radius of Earth's orbit subtends an arc-second (one sixtieth of one sixtieth of a degree). A parsec is about 3.2 light-years. To set the scale, the distance from the Sun

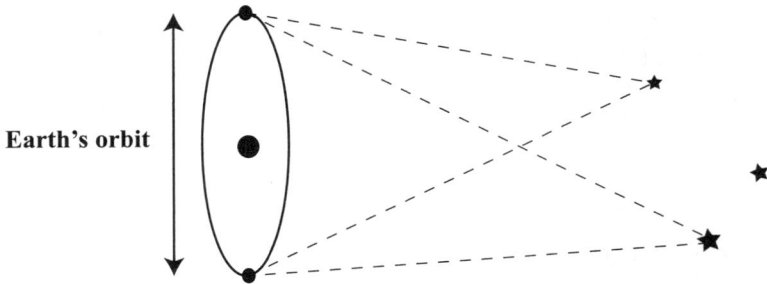

FIGURE 7.6 Using the Earth's orbit as a baseline for measuring
the distance to nearby stars using parallax.

to Earth is eight light-minutes (150 million kilometers, 100 million
miles). The distance to the nearest star, Proxima Centauri, is four
light-years. The Earth–Sun distance compared to the Sun–Proxima
distance is like looking at a pea from a mile away.

DETERMINISM

The notion of a clockwork universe naturally leads to the idea that
if all the motions of all the particles of the universe were known,
and the laws of motion, such as gravity, were known that all future
events could be predicted with infinite precision. This was famously
articulated by the mathematician Pierre-Simon Laplace, who wrote
in "A Philosophical Essay on Probabilities" (1814):

> We may regard the present state of the universe as the effect of its past
> and the cause of its future. An intellect which at a certain moment
> would know all forces that set nature in motion, and all positions
> of all items of which nature is composed, if this intellect were also
> vast enough to submit these data to analysis, it would embrace in a
> single formula the movements of the greatest bodies of the universe
> and those of the tiniest atom; for such an intellect nothing would be
> uncertain and the future just like the past would be present before
> its eyes.[6]

The "intellect" in his essay became known as Laplace's demon. This concept raises an interesting question about free will. We feel that we have freedom to make decisions, but if this viewpoint were correct, then it implies that we do not really possess it, and the concept of decision-making is just a kind of hallucination or mirage that gives the appearance of free will.

This is not necessarily a new idea, as a similar concept was articulated by Cicero in his work *On Divination*:

> Moreover, since, as will be shown elsewhere, all things happen by fate, if there were a man whose soul could discern the links that join each cause with every other cause, then surely he would never be mistaken in any prediction he might make. For he who knows the causes of future events necessarily knows what every future event will be. But since such knowledge is possible only to a god, it is left to man to presage the future by means of certain signs which indicate what will follow them.[7]

What was new with Newton's *Principia* is that it seems to give a sort of portal to how an intellect might carry out the calculations necessary to play forward all the motions of all the particles.

For a Laplace-style demon computation, there are problems. How would the computer compute itself, for example? But even beyond this, to do computations you must break the computations into discrete steps in space and time. Once you get discrete steps in space and time, you then get into the problem of unpredictable chaos. Here the sense of a clockwork universe seems to break down.

In fluid dynamics, for example, there are certain regimes where molecules follow well-described predictable paths, called laminar flow. Elaborating, we sometimes speak of linear and nonlinear systems in physics. Imagine a microphone, an amplifier, and a speaker. The microphone might increase the loudness of a voice by a factor of 10. If I speak at a "one," I get a loudness of 10 out of the speaker. If I speak at a "two," I get a loudness of 20 out of the speaker, and so forth. The loudness out of the speaker is linear with the loudness of the voice. On the other hand, if I bring the microphone close to the speaker, a tiny noise from the speaker gets picked up by the microphone, which gets amplified, and the microphone picks up this amplified sound. This creates a runaway feedback effect that results in a screeching sound. We would say that this puts the speaker+microphone system into a nonlinear regime.

Returning to fluid flow, the equations for fluid flow are classical, based on Newton's laws governing objects that accelerate, but they do contain terms that aren't linear, like a position or velocity squared. When the nonlinear terms are small, fluid flow is predictable and can be calculated, but when the nonlinear terms get large, turbulent flow sets in. In this case, the flow is unpredictable and vortices and sorts of strange unpredictable effects begin to set in (see fig. 7.7).

The equation governing fluid dynamics is called the Navier-Stokes equation. There is no full solution to the equation, only approximations. In fact, no one has been able to solve for fluid flow fully

laminar flow

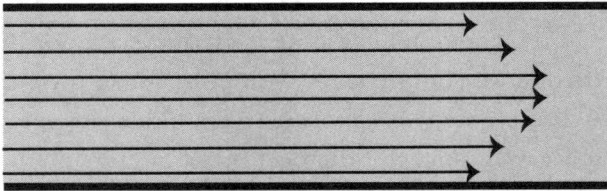

turbulent flow

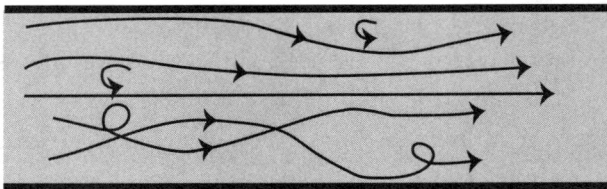

FIGURE 7.7 Laminar and turbulent fluid flow through a pipe.

on a computer—again, only approximations. The Clay Institute of Mathematics has offered one of its Millennium Prizes for a proof of the existence of solutions.

Weather forecasting has this problem in spades. Air is a fluid, and air flow is described by the equations describing fluids. In addition, there is the question of water going into solution in air, which absorbs heat, or condensing, which releases heat. The combination of the fluid flow and water condensing makes weather forecasting difficult, particularly on long timescales. Currently, computer-generated forecasts over timescales of about five days are fairly accurate but beyond this they become far less reliable. Even more confounding are intense short-distance scale storms, like afternoon thunderstorms and hurricanes, which are highly unpredictable.

One computational difficulty is that a very small difference in initial conditions can give rise to divergent behaviors. Edward Lorenz was an MIT mathematician and meteorologist who studied the numerical computation of equations like the ones governing weather. Lorenz found that even very tiny differences in the starting conditions for a computer simulation would give rise to vastly different final results. This became known as the "butterfly effect." A famous quote of his is, "Does the flap of a butterfly's wings in Brazil set off a tornado in Texas?"

A similar issue is called the three-body problem. In Newton's gravity straightforward solutions exist for a two-object (body) system. In our solar system the Sun's gravitational pull is vastly stronger than the interactions among planets so, effectively, planetary motion can be broken down to a series of two-body systems between each planet and the Sun, making calculations straightforward. The influence of planets on each other can then be performed as a kind of second approximation. When three bodies are involved generally, it's no longer possible to write down a full solution, and the motions can be quite chaotic, depending on the masses and initial conditions.

The Earth–Moon–Sun system is a three-body system, but the influence of the Earth and Moon on the Sun are negligible.

Nonetheless, calculating the Moon's orbit proved to be a challenge due to the Earth and Sun both exerting a gravitational force on the Moon, and it wasn't until the latter half of the eighteenth century that it was solved.

More generally, three (and more) body problems can be quite chaotic in motion. This notion formed the basis of Liu Cixin's science fiction novel *The Three-Body Problem*, where an extraterrestrial civilization at the nearest star system, Centauri, exists on a planet that has three stars orbiting each other. In the novel the development of physics on the distant planet is delayed because the star system exhibits difficult behaviors to calculate, while our solar system is dominated by the Sun, and planetary motions can be described as a set of two-body motions.

WILLIAM BLAKE'S REACTION TO NEWTON

The work of Galileo, Kepler, and Newton represented a breakthrough in our understanding of motions on Earth and in space, with many follow-ons, like fluid dynamics. But there was also a cultural reaction against the new physics. Many saw it as taking away a sense of mysticism or wonder at the world, reducing it to a cold, stark set of equations. One notable example was artist and poet William Blake, who was deeply opposed to the Enlightenment. In particular, he saw Newton as banishing God and replacing him with a mechanistic universe that was nothing but cogs in a giant, heartless machine. Blake was quoted as saying, "Art is the tree of life, science is the tree of death."

One of Blake's most famous works is *Newton*, a print showing him sitting on a rock covered in sea-life, evidently at the bottom of the sea (fig. 7.8). Newton is depicted as naked, crouching over a scroll and somehow deeply engaged in a drawing using a drafting compass.

Blake also wrote about Newton in a couple of poems. In "You Don't Believe," Blake wrote:

FIGURE 7.8 The painting *Newton* by William Blake.

You don't believe—I won't attempt to make ye.
You are asleep—I won't attempt to wake ye.
Sleep on, sleep on, while in your pleasant dreams
Of reason you may drink of life's clear streams
Reason and Newton, they are quite two things,
For so the swallow and the sparrow sings.
Reason says 'Miracle,' Newton says 'Doubt.'
Aye, that's the way to make all Nature out:
Doubt, doubt, and don't believe without experiment
That is the very thing that Jesus meant
When he said: "Only believe." Believe and try,
Try, try, and never mind the reason why.

Likewise, poet John Keats was no fan of Newton, much in the same spirit as Blake. He is said to have often proposed a toast, "Confusion to the memory of Newton."

ABSOLUTE TIME

The Newton/Galileo version of space ruled the stage for the development of physics and astronomy for 218 years and beyond. This includes formulations of electromagnetism, fluid motion, and celestial motion. Many current developments still use this formulation because it works in many domains where velocities are much less than lightspeed.

A cornerstone of the Newtonian/Galilean space is the concept of absolute time, where all clocks tick at the same rate for all observers. This seems like common sense because this is consistent with our experience. We can synchronize watches and go about our business believing that they'll always tell the same time. This starts to break down when relative velocities start to become a significant fraction of the speed of light. In these conditions we can no longer rely on an absolute sense of time, and time itself gets mixed into our models of space. This was a key development that came to Einstein in an epiphany.

THE WEDDING *of* SPACE
and TIME : RELATIVITY

"Absolute space" and "absolute time" were two terms used by Newton as a foundation for his theories of motion and dynamics. In many ways these terms might make sense to us Earth-bound creatures. Many of us desire a fixed scaffolding to define our space: countries, lots, villages, and dwellings. Likewise, while we don't like to be ruled by the clock, we all use a commonly understood sense of time to synchronize our events. It seems reasonable that we would use generalizations of these and apply them to the universe. But are space and time really absolute? This is what Albert Einstein contemplated while trying to make sense of the emerging theory of electric and magnetic fields.

In a journey to the roots of science Einstein realized, in a sudden flash of insight, that space and time were not absolute but tied together as a single entity: "space-time." Not only this, but space-time will vary, depending on the relative motion of two objects. Einstein published his account of this special theory of relativity in 1905 and in 1913 he published a general theory of relativity, where he proposed that a curvature of space-time is associated with gravity.

Not only did the emergent theory of relativity shake science to its core, it also spawned cultural manifestations, where all things *seemed* to be relative, including truth. It is even said that the modern era started with the advent of relativity.[1] Before turning to general relativity, I'll first start out with special relativity and some cultural manifestations, followed by the general theory.

FRAMES OF REFERENCE AND RELATIVITY

My mother is an amateur fan of science but has always been perplexed by the term "space-time." I set out to try to explain it for her in writing what follows. But, alas, I fear that it's still elusive for her. The following descriptions of observers, clocks, and yardsticks may be a bit too much for some readers. For this I apologize, but I include it because I think some readers may be able to follow the discussion and perhaps grasp the concept of "space-time," and why we speak of it as one entity.

We touched on frames of reference in the previous chapter regarding how an event might be viewed from the perspectives of two trains and a train station platform. To understand Einstein's relativity, we might endeavor to think like him, which I realize is a tall order, but I will try to walk the reader through it.

Einstein was influenced by the philosophy of David Hume, and physicist/philosopher Ernst Mach. For Mach, reality is embodied by the possibility of linkages of sense experiences and measurements as the tangible things in this world. Only relations among the measurements we could carry out have real meaning. We can phrase these relations in terms of observers who can carry out measurements, but this doesn't imply that there must be observers. What counts is the possibility of making observations.

We first need to understand space and time from the perspective of Galileo's and Newton's construction before proceeding to Einstein's space-time.

Returning to the concept of a frame of reference. We can think of it as an abstraction of constructs you encounter in everyday life. One example might be a town with a regular grid of streets on it. Say the grid has the labels "A Street," "B Street," . . . with crossroads "First Avenue," "Second Avenue," and so forth. If you wanted to meet a friend, you can use this naming convention to say, "Let's meet at the corner of D Street and Third Avenue at 4 PM tomorrow." This is sufficient to establish the meeting place and time unambiguously. Let's take another example: a moving airplane. All the seats are laid out on a grid and although they're moving at a high speed with respect to the ground the seats keep their positions with respect to each other. A flight attendant might call the pilot and say, "The passenger in seat 24C said something rude to me three minutes ago." As with the time of the meeting at D and Third, this establishes an unambiguous location and time of the rude comment.

You can think of a frame of reference as an abstraction of the above concepts. The idea is that you can lay out a grid called a coordinate system that specifies locations in space and a time that are unique to that frame of reference. Multiple frames of reference can move at speeds relative to one another and, in principle, there can be observers occupying the frames of reference who can compare notes with each other. The observers can compare notes both within their own frame of reference and with observers in the frame of reference moving relative to them. As an example you may be driving with friends through a town—people on the street witness a bank robbery at the corner of Sixth and B, and as you drive by at just the time of the robbery you take a look at your watch and note this. Later the police ask for details—both you and your friends in the car, and the people on the street can agree that the robber left the bank at 10:15 at the corner of Sixth and B.

Now this may all seem trivial as it matches our common experience, but just to hammer it home let us look at a frame of reference as a kind of scaffolding that has unique positions everywhere in space and at each of those positions there is a clock that registers time in that frame. The clocks all keep the same time—so 10:15 at Sixth and B is also 10:15 at Third and D. Likewise, 10:15 at seat 24C on the airplane is the same as 10:15 in the cockpit of the airplane, and even 10:15 on the ground at Third and D. At least this is what we commonly think and is how Newton's laws work.

As mentioned in the previous chapter, there is a transformation that allows us to move from one frame of reference to another. Formally, a transformation can be expressed as a mathematical relation between the coordinates and times of the two frames, but I'll spare the reader these details. The importance of a transformation is that the laws of nature should not change from one frame of reference to the next. This is the principle of relativity, both for Galileo/Newton and Einstein. The form of the transformation of Galileo and Newton was amended by Einstein to be more universal.

Using transformations, we can relate events in two frames of reference when they're moving with respect to each other. In the example below, I'll use a conference table with two people named Sam and Jess, both seated at a table. Their seat positions represent their spatial coordinates, and they have clocks that they hold in their hands and can consult.

Now, let us imagine there are twins of Sam and Jess, also at a conference room table, but one that is moving with a speed relative to the original in figure 8.1. The twins have the same respective positions at

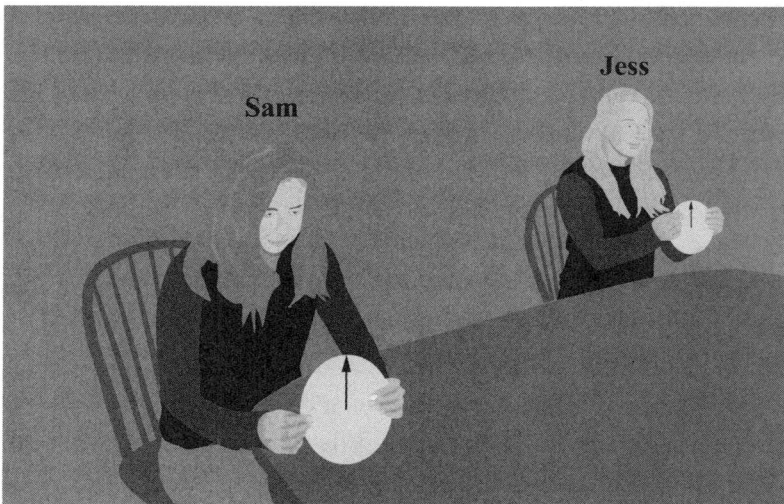

FIGURE 8.1 One reference frame, where Sam and Jess occupy two different seats at the conference table and both have clocks.

FIGURE 8.2 Twins of Sam and Jess in a reference frame
that is moving with respect to the other frame.

the conference room tables as the non-twins and have synchronized clocks (see fig. 8.2).

Now, say that the twins on the moving frame can pass seamlessly through the others and each can spy on the others' clocks. In figure 8.3 I show Sam and Jess and their twins at some moment as they are moving past each other. With the relative motion of the two conference tables, what they see in the other frame changes with time. Because the twins are moving from right to left, at an early time, Sam's twin overlaps with Jess, and at a later time, Sam's twin overlaps with Sam.

The two frames passing through each other is illustrated in figures 8.3 and 8.4. Figure 8.3 shows an earlier time when Sam's twin is halfway between Sam and Jess. Figure 8.4 shows a later moment when Sam and her twin overlap. For simplicity, I only show the clocks of the twins in the latter.

Now, one important feature of the Galilean transformation above is that *all* the clocks in *both* frames of reference register the same times when they are compared. The comparisons depend on

FIGURE 8.3 A snapshot in time where Sam, Jess, and
their twins pass through each other.

FIGURE 8.4 A later moment when Sam, Jess, and their respective
twins overlap. In this figure I only show the clocks of the twins.

where they overlap, but this is not terribly significant as the clocks all tell the same time.

Sam and Jess can also figure out how fast their twins are moving. Distance traveled divided by the time difference gives us a speed. If Sam and Jess want to find the speed, they can look at the time when Sam's twin overlaps with Jess, and then the time when Sam and her twin overlap, and use the distance between the chairs divided by the time interval to find the relative speed.

With the above construction in mind, we can turn to the concept of "absolute time." There is an absolute time built into the transformation. All clocks run at the same rate, and all comparisons between clocks at different points of overlap will show an agreement between the two frames.

Newton's laws do not privilege any particular frame of reference. Although Newton has used the term *absolute space*, there is nothing built into Newton's laws of motion that requires this. There is, however, a different kind of "absolute" in Newton's formulation—namely that the distances between objects or events are the same in different frames of reference. Said another way, if Sam and Jess are two meters apart in their frame of reference, their twins will measure their separation to be two meters apart as they pass by in the moving frame of reference. In this sense there is a kind of absolute space, where distance intervals are the same for an object that is stationary with respect to us, or if it's moving.

What is relative about the above formulation? From the point of view of Sam and Jess, their twins are moving from *their left to their right* at a certain speed. From the point of view of the twins, Sam and Jess seem to be moving from *their right to their left*, with the same speed, but in the opposite direction.

The principle of "relativity" means that the laws of physics do not distinguish among any frames of reference. This means that forces that appear in one frame will act the same as forces that appear in another frame. Constants of nature, like the strength of gravity, or the charge on an electron, will be identical in all frames. This was already a principle for the physics of Newton, and this "invariance" is based on the Galilean transformation above.

I will add a couple of ideas that you might have some intuition about. Let's imagine that someone lights two firecrackers at either end of the conference table that go off at the moment when Sam, Jess, and their twins overlap. They'll all agree on when it happens because they can read that moment from their clocks. They can also agree on the distance between the two firecrackers by observing the positions of the bangs. The idea that there is an agreement about the separation between the two events in the two frames is called *simultaneity*, meaning that if the firecracker bangs happen at noon sharp in the Sam/Jess frame, it happens at noon sharp in the frame of the twins.

Another idea that seems to be common sense is the way velocities add together. We can compare what a person on a train platform sees with respect to a person on a train. Say the train is moving at 40 miles per hour, and a person on the train throws a ball forward in the aisle at 20 miles per hour; a person on the train platform will see the ball moving at 60 miles per hour.

Finally, I need to point out one critical point again: these constructions invoke the use of observers and clocks, which are not necessary, but are anthropomorphic constructions that aid the reader to understand the constructions. There is a kind of egocentrism of the constructions in the sense that we invoke the presence of human observers to explain and understand the transformations.

SPECIAL RELATIVITY

Why modify a model of space and time that agrees with our intuition? This brings us to Albert Einstein's background.

In the latter half of the nineteenth century much progress was made understanding electricity and magnetism. While for decades they were viewed as two different phenomena, it was ultimately realized that they were two manifestations of the same thing: an *electromagnetic* field.

Albert Einstein's father and uncle were in the electrical business, manufacturing lamps, transmission systems, and other devices. When he was young, he got exposure to all manner of electrical devices and learned their workings. To Einstein there appeared to be paradoxes that arose in certain experiments involving the relative motion of magnets and wires.

Another, more important issue had to do with how light travels through space. The unification of electricity and magnetism showed that light is a vibrating electric and magnetic field that travels through space. The speed of light, already measured in several experiments, itself emerges from the equations for electromagnetism. When certain constants of nature associated with electricity and magnetism are combined, the speed of light naturally arises, and appears to be a constant of nature in the same way the strength of Newton's gravity is a constant of nature.

Light was already known to be a wave for quite some time through numerous experiments. In the nineteenth century, when the theory of electricity and magnetism were being formulated, waves were thought to be a disturbance in a medium. Sound waves are vibrations of air molecules where a disturbance moves through the atmosphere, but the air doesn't come along with the disturbance. Likewise, ocean waves are a disturbance in the surface of the ocean, but the ocean itself is not moving along with the wave.

With this line of reasoning, light waves would be a disturbance in some medium. But what medium?

Light travels through the vacuum of outer space. We see it reaching us from the Sun, Moon, planets, and distant stars. If there was a medium out there, it would have to permeate all outer space and be unlike any substance we know on Earth. Ideas of something other than earthly matter had been around for quite some time. Aristotle talked of a substance called *ether* that is lighter than air and floats above the Earth as a kind of fifth element. Ideas circulated about an ether permeating all of space emerged in the nineteenth century to describe how light waves could propagate.

As the electromagnetic theory matured, many scientists felt that they could detect this mysterious ether by measuring the speed of the

Earth moving through it. In some ways, the existence of ether in outer space indeed implied a kind of absolute space defined by the frame of reference of the ether. Attempts to measure the motion of the Earth through the ether came up empty-handed. There were attempts to patch up the ether theory, however. One thought was that the motion of the Earth pulled along the ether as a kind of cloud following it. This was called ether drag but attempts to find ether drag fizzled. It seemed that the speed of light was a constant and didn't depend on the motion of the Earth through any kind of fixed medium.

There was another approach, that time passed more slowly in a frame moving relative to another frame, and that distances were compressed. The changes in time and space happen in just such a manner as to explain the constancy of the speed of light. This approach was advocated by two physicists, Hendrik Lorentz and Henri Poincaré. Although this patch "worked" in some sense, it was a mathematical sleight-of-hand, and did not explain what was going on in a deep way.

Here is where we pick up the story of Einstein. He was aware of the above ideas about light but found them unconvincing. In addition, he was very keen on a deeper understanding of the workings of nature rooted in philosophy. As sometimes happens, there are serendipitous moments in history, and the chance meeting of Einstein with a man named Michele Besso was such an event. When they first met in Zurich, Besso was about twenty-three years old, and Einstein seventeen. They played music together. Besso and Einstein's friendship developed and centered around long conversations about philosophy and other matters.

Besso introduced Einstein to the writings of physicist/philosopher Ernst Mach—in particular Mach's book, *The Science of Mechanics*. Mach's view of the world was that the only reality was events that could be linked together. For example, when a clock's minute hand reaches 12, coinciding with the arrival of a train, the two events are linked. Mach also took a dim view of the underpinnings of the Newtonian universe—with the "conceptual monstrosity of absolute space. . . . purely a thought-thing that cannot be pointed to in experience."[2]

The importance of the philosophical process for Einstein during this period of his life cannot be overstated. Here, I quote Gerald Holton, who has written extensively about Einstein:

> The philosophical worldview of a scientist surely is as important as, for example, his understanding of the mathematical tools of the trade. I am thinking here of how Einstein confessed to having been influenced by his reading of David Hume and Ernst Mach.[3]

One example of Hume's influence on Einstein is a quote from *The Treatise on Human Nature*:

> Habit may lead us to belief and expectation, but not to knowledge, and still less to the understanding of lawful relations.

This quote, a favorite of Einstein's, shows how he was digging deep to find something more fundamental than the patches proposed by Lorentz and Poincaré.

With this in the background, Einstein pondered the theory that unifies electricity and magnetism. The speed of light emerged as a constant from the equations. There is no mention of ether—there is no invocation of rulers contracting or clocks ticking at different rates. Einstein was convinced that since the speed of light was something so fundamental to the theory of electromagnetism it must be a constant of nature. The problem is that the existence of an ether created a preferred frame of reference—an absolute space. This cuts against the grain of the idea that the laws of physics should not change for observers in different reference frames. For the Galilean transformation articulated above, if you shine a flashlight on a moving train and compare the results of a measurement on the train to an observer on the platform, you might expect to obtain different values, and this would imply that the speed of light was not a constant of nature.

Rather than being satisfied with the ad hoc solution proposed by Lorentz and Poincaré, Einstein dug deeper, pushed on by his philosophical discussions with Besso. Einstein looked at the problem of observers in two frames of reference as above: the frames are a kind of scaffolding where observers can sit, make measurements, and consult clocks. Again, these are not real scaffolds, observers, or clocks, but are a way to "construct" the rules for the transformation between frames.

The concepts of relativity can be difficult to grasp because it seems counterintuitive for many, and explanations to interested lay persons often come up short. For example, here is a snippet from a *New York Times* editorial published in 1928, "A Mystic Universe," where the author laments the difficulty grasping the tenets:

> Countless textbooks on Relativity have made a brave try at explaining and have succeeded, at most, in conveying a vague sense of analogy or metaphor, dimly perceptible while one follows the argument painfully word by word and lost when one lifts his mind from the text. It is a rare exposition of Relativity that does not find it necessary to warn the reader that here and here and here he had better not try to understand. Understanding the new physics is like the new physical universe itself. We cannot grasp it by sequential thinking. We can only hope for dim enlightenment.[4]

One of the obstacles to our intuition about space-time is that Einstein's special relatively is only necessary when objects are moving close to the speed of light: 186,000 miles per second. Because none of our daily interactions deal with that kind of speed, our brains aren't wired to grasp the concepts. At some level, the physics from 1900 onward progresses into domains of space, time, and speeds where the human mind had not evolved to grasp, and as we move forward, we will see more concepts of space that can be difficult to make sense of.

The turning point for Einstein was a conversation with Besso one Saturday night in Bern, Switzerland. Einstein announced to Besso, "Recently I have been working on a difficult problem. I have come here to do battle with you." They discussed the ideas of Ernst Mach and how there is no absolute motion, only relative motion as seen by different observers. Evidently, this gave the spark of a new idea for Einstein, who told Besso the next day that he had solved the problem.[5]

What was this sudden insight? Recall that I mentioned the two firecrackers going off at the same time on the ends of the tables for Sam, Jess, and their twins. These events are simultaneous, and the concept of simultaneity in all reference frames is something we just assumed to be true. After all, it seems to be consistent with our experience. Einstein realized that he could solve the question of the constancy of the speed of light if he gave up on the concept of simultaneity. There is nothing carved in stone that mandates that if two events are simultaneous in one frame of reference, they must be simultaneous in another. The idea of simultaneity comes from our intuition based on experience with relative speeds much less than the speed of light—but if this gets modified at speeds approaching that of light, the contradiction can be solved. Dropping simultaneity gives the freedom to change the above transformation construct—in particular by allowing the clocks (really the time) in one frame to be dependent on position when viewed from another frame moving relative to it. This implies that the spatial coordinate in the direction of motion and the time get mixed together: one can no longer talk about an absolute time, but rather a "space-time." In his 1905 paper on special relativity Einstein wrote about this:

> If we wish to describe the motion of a material point, we give the values of its co-ordinates [spatial location] as functions of the time. Now we must bear carefully in mind that a mathematical description of this kind has no physical meaning unless we are quite clear as to what we understand by "time." We have to take into account that all our judgements in which time plays a part are always judgements of simultaneous events. If for instance, I say, "That train arrives here at 7 o'clock," I mean something like this: "The pointing of the small hand of my watch to 7 and the arrival of the train are simultaneous events."

It might appear possible to overcome all the difficulties attending the definition of "time" by substituting "the position of the small hand of my watch" for "time." And in fact such a definition is satisfactory when we are concerned with defining a time exclusively for the place where the watch is located; but it is no longer satisfactory when we have to connect in time series of events occurring in different places, or—what comes to the same thing—to evaluate the times of events occurring at places remote from the watch.[6]

There is some echo of Hume's pronouncements on the nature of time:

The idea of time is not derived from a particular impression mixed up with others, and plainly distinguishable from them; but arises altogether from the manner, in which impressions appear to the mind, without making one of the number.[7]

The above snippet from his paper on special relativity reflects the influence of Mach and Hume on Einstein's thinking, and his late-night battle with Michele Besso. History often seems to proceed like the stick-slip model of earthquakes: strain builds between rocks on either side of the fault line for a long period. The rocks may bend, but they do not give. Suddenly the pressure is released and the land on either side of the fault line shifts a significant distance, radiating energy throughout the Earth. As it is with stick-slip earthquakes, so it was with the advent of special relativity. The Newtonian concepts were under significant stress with accumulating evidence. Then one evening a sudden epiphany struck Einstein, who articulated it in his *Annalen der Physik* paper quoted above. The seismic shift changed physics forever. Having slipped the bondage of simultaneity, Einstein had the freedom to create a transformation law that preserved the speed of light as a constant for observers in different frames of reference.

How does this work in practice? Let's go back to the arrangement of Sam, Jess, and their twins. We have four possible cases of measurements for the two frames of reference moving with respect to each other:

1. Sam and Jess can measure the speed of light and get a value.
2. Their twins can measure the speed of light and get the same value.
3. Sam and Jess can see their twins measure the speed of light and they would see their twins derive the same value.
4. The twins can look at Sam and Jess measure the speed of light and see them derive the same value.

All of these cases have to work. In addition, the concept of a relative speed between the two frames must have the same meaning as before. That is to say, one can measure the relative speeds of the two frames by taking a distance divided by a time measured in one frame for the other.

When Sam and Jess looked at their twins in their frame, the clocks were synchronized in both frames of reference. If we abandon absolute time, the idea that the clocks are synchronized, and allow them to vary with position from one frame to the next, we can create a transformation rule that preserves the speed of light. In other words, events that occur simultaneously in one frame do not appear simultaneous in another frame. In addition to this, the speed of clocks moves differently in the two frames, and distances change depending on the relative speeds of the two frames. Figure 8.5 is the analog of figure 8.3, except it is now changed for the effects of special relativity. This is seen in the frame of Sam and Jess, where the positions of the twins and the clocks are visible. Figure 8.5 is drawn from the point of view of Sam and Jess's frame and shows what they see in the twins' frame.

The first thing to note in figure 8.5 is that, while the times on the clocks of Sam and Jess are synchronized, what they see on the twins'

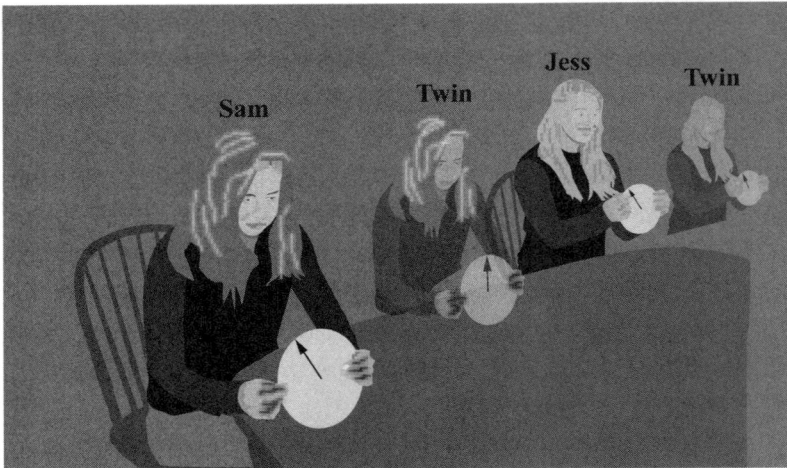

FIGURE 8.5 A comparison between the frames of Sam and Jess when their twins are moving at speeds approaching the speed of light. The clocks of the twins are no longer synchronized as seen from Sam and Jess's frame.

clocks are (1) different from theirs, and (2) different between the two twins. In other words, the times on the twins' clocks that they see are different and depend on the location of the twin.

Figure 8.6, the analog of figure 8.4, shows an image of what happens a moment after the scene in figure 8.5. In it you can see that when the position of Sam and her twin line up, the position of Jess and her twin doesn't line up. In addition, the rate at which the clocks tick is different for the two twins and depends on their location.

This is what "space-time" means—in the first example of the Galilean transformation, we could speak separately about space and time. The clocks would go about their business—everything happily synchronized, and the positions would go about their business. But in special relativity

1. Clocks move at different speeds in different frames
2. Spatial intervals change when comparing one frame to another
3. Time becomes a function of position when one frame is viewed from another

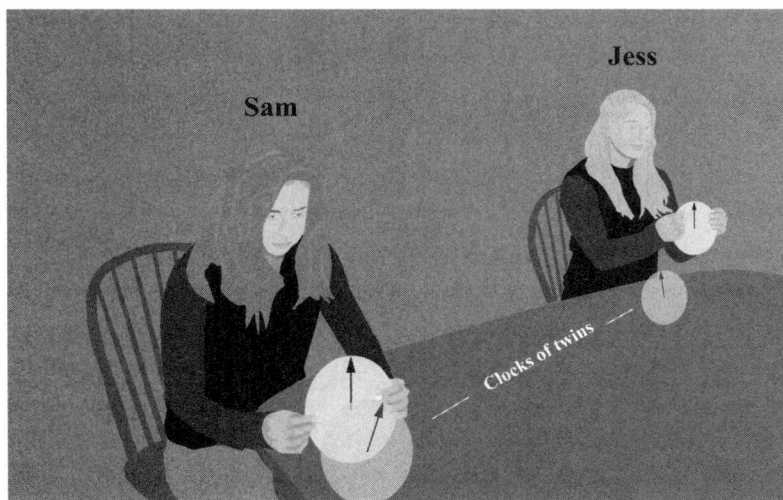

FIGURE 8.6 When Sam's clock is coincident with her twin's clock, they don't show the same time. In addition, Jess's clock now does not line up with her twin's, as the spatial intervals have changed as well.

The above issue of clocks when viewed from one frame to another might seem random at first glance, but it isn't. There are two constraints that Einstein applied:

1. Velocity retains its meaning of distance divided by time
2. The speed of light is constant for all observers in all frames

These constraints, coupled with the idea that clock time can be a function of position, is enough to create the transformations of special relativity that are called Lorentz transformations. This is in distinction to the first transformation law above (Galilean transformation) that had the idea of simultaneity built into it. Simultaneity gets traded for the constancy of the speed of light.

To amplify on the discussion, figures 8.7, 8.8, and 8.9 show two frames of reference, A and B. When both are at rest relative to each other, they have the same structure, with four clocks located at equal distances from each other. When at rest (fig. 8.6) all the clocks are

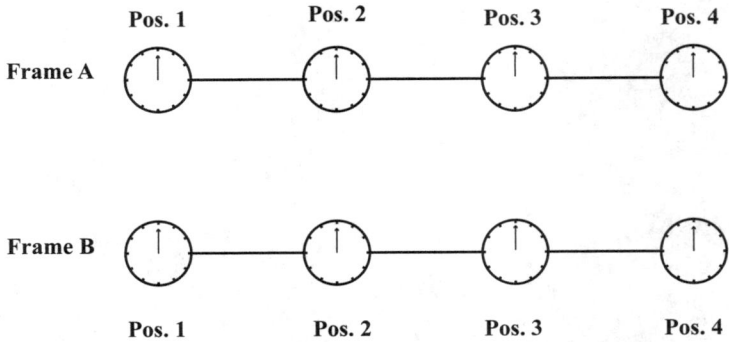

FIGURE 8.7 Frames A and B when they are at rest with respect to each other.

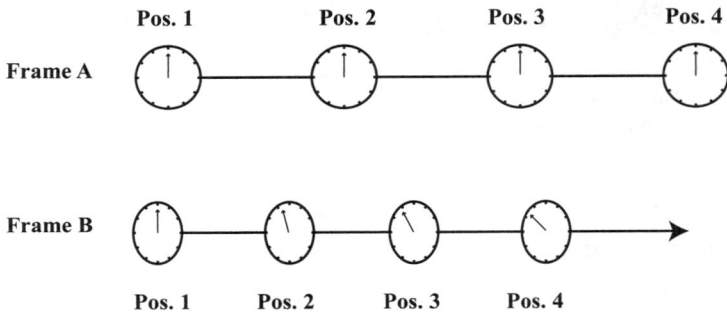

FIGURE 8.8 Frame B moves to the right at half the speed of light relative to frame A. This is what observers in frame A see for the clocks and positions of the equivalent locations in frame B as it passes by.

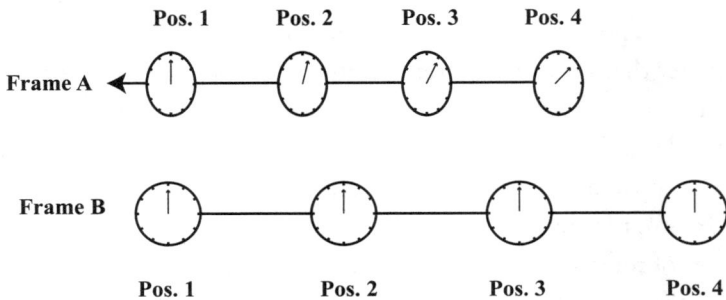

FIGURE 8.9 Frame A as viewed by frame B moves in the opposite direction with the same speed. This is what a viewer in frame B will see for the clocks and positions of equivalent locations.

synchronized in both frames, and observers stationed at each clock will see the same location opposite them at the same time. I label the positions as Pos. 1, Pos. 2, etc.

Now, I take the case where the two frames are moving relative to each other and ask what they look like. Figure 8.8 shows what frame B looks like from the viewpoint of observers in frame A. I have chosen to have the clocks at position 1 synchronized in both frames. This can only be done at one location for one time. Distances are contracted in frame B and the clocks move at different speeds. In addition, the times on the clocks in frame B no longer appear synchronized from the standpoint of observers in frame A. The clock times vary with the position seen in frame B. I can reverse the views and ask what observers in frame B see when looking at frame A (see fig. 8.9). Again, the distances between the observation positions appear to shrink, and the times on the clocks depend on which positions are being observed from one frame to the other.

One important point is that the old Galilean relativity still exists as a limit when velocities are small compared to the speed of light. In this sense, Einstein's relativity doesn't really "smash" or "destroy" the old Newtonian concepts. Special relativity has Newton's concepts as a limiting case.

Another important point has to do with the word "absolute." In a sense, Einstein traded one absolute for another. Absolute space and time had to be abandoned in order to get an absolute speed of light.

It's important to note that the transformation laws Einstein articulated are mathematical. To save the reader the agony of the math, I have tried to use words to describe what happens, but at the end of the day, it's really the transformation rules that give a full description.

There are a large number of predictions from special relativity. Two significant effects are worth noting:

1. Time appears to move more slowly in fast moving objects
2. Objects shrink in the direction of motion.

THE TRAIN PARADOX

There are some measurements that seem like a paradox when first considered, but you have to be careful to specify how they are carried out. Consider this: if an object seems to shrink in the direction of motion for one observer, what does it look like when you're in the other frame of reference?

At first glance it doesn't make sense. A rod in frame A looks shortened in frame B, but then a rod in frame B would look shortened in frame A.

It's not a paradox. The following example shows how it works once we abandon simultaneity. Consider the following situation—on the distant planet Caladan scientists have figured out how to get trains to move close to the speed of light. They decide to do an experiment. There is a mountain that has a tunnel going through it that is shorter than the train when the train is at rest. This is shown in figure 8.10.

When the train moves close to the speed of light relative to the ground, it contracts in the direction of motion. But for an observer on the train it would seem that the mountain shrinks, as it seems to be moving backward relative to the train. Which one really shrinks? The scientists on Caladan decide to "capture" the train by installing doors on either end of the tunnel. One group of scientists is stationed

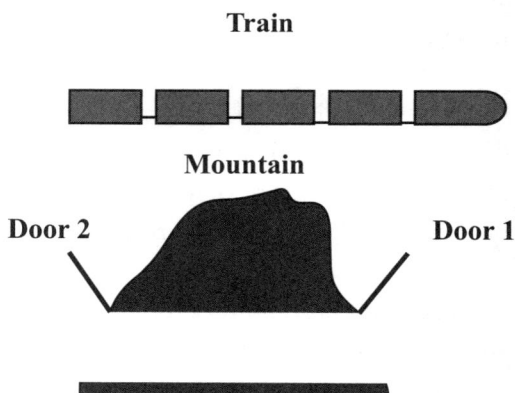

Train

Mountain

Door 2

Door 1

FIGURE 8.10 Train and tunnel both at rest on the planet Caladan.

at either end of the tunnel with the doors. Another group boards the train to see what happens from the train's frame of reference.

The train gets moving and the scientists on the ground close the doors on both ends of the tunnel when the contracted train is inside, and then immediately open up the doors so the train can exit without an accident. The train is "captured" (see fig. 8.11). You can see that the train has shrunk in the direction of motion so that it seems to fit into the tunnel, and the scientists on the ground can close both doors simultaneously, and then immediately open them, lest the train smash into door 1.

What about the view for the observers on the train? It seems like a paradox, but in special relativity we have let go of the idea of simultaneity. The observers on the train see that the mountain is moving backwards close to the speed of light and that it's contracted. The train clearly seems to be longer than the tunnel. As the train enters the tunnel, the scientists on the train see the far door first (door 1 in fig. 8.12) close and then open. As the front of the train leaves the tunnel, eventually the rear of the train passes the rear entrance. Only then does the other door (door 2 in fig. 8.13) close and then reopen.

So, from the point of view of the scientists on the train, the doors don't close at the same time—first one and then the other closes. But, from the point of view of the scientists on the ground, they close at the same time.

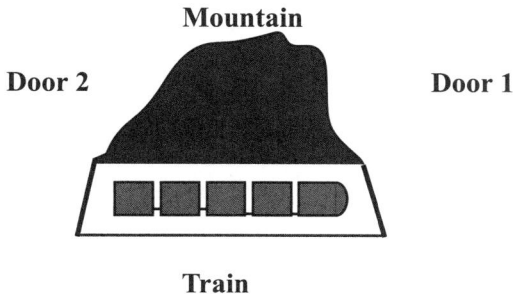

FIGURE 8.11 Moving train at its moment of capture in the tunnel as seen from the frame of reference of the ground. The scientists "capture" the train by closing both doors simultaneously.

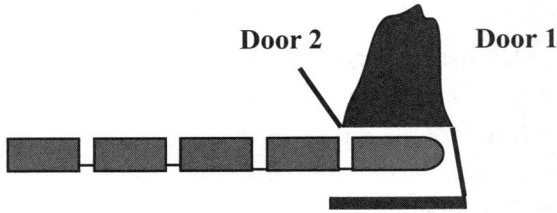

FIGURE 8.12 In the frame of reference of the train, the far door (door 1) closes first, as the train enters the tunnel, but door 2 remains open.

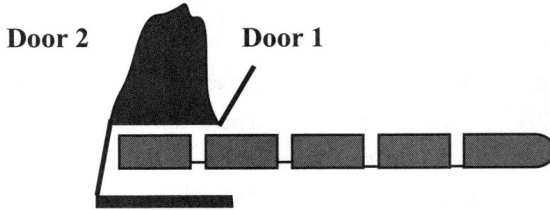

FIGURE 8.13 As the train exits the tunnel, door 1 opens, but the other door (door 2) is closed.

The resolution to this paradox has everything to do with how measurements are carried out. A number of other apparent paradoxes are created by special relativity, but they can all be resolved by carefully specifying how the measurements are performed.

Constructions like the above train and tunnel can also be set up to demonstrate that observers in different frames of reference can measure identical values for the speed of light and verify that the measurements performed in another frame yield an identical value. This was, in fact, the very basis for Einstein's special relativity, but it is still reassuring that it emerges when the experiments are carried out carefully.

While the effects of special relativity might seem rare, there are definitely measurable consequences that have been verified in a vast array of experiments. It is most evident with subatomic particles that we can accelerate to speeds close to the speed of light. Some particles will decay into lighter ones with a known decay rate. The decay acts like an internal "clock." We can measure the decay rate when the particle is at rest and get a value. If the particle is accelerated to a high speed—close to the speed of light—its decay is slowed down. All modern particle physics theories, including the highly successful Standard Model of particle physics, have special relativity as a bedrock. For high-energy accelerators to work, we absolutely must design them with the effects of special relativity included.

Even technologies we now consider every day use special relativity. One example is global positioning systems (GPS). There are dozens of GPS satellites orbiting the Earth at any moment. Each satellite carries an atomic clock that has a precision of one billionth of a second (a nanosecond). They beam out information that gives a known satellite location and time stamp. On Earth our GPS units receive this beamed information and can see four or more satellites at once. The receivers then calculate a position with a precision of roughly a meter by combining the beams from the satellites. To get this accuracy, the clock time on the satellite must be known to about 20 nanoseconds.

Now, satellites are moving relative to the ground at 7,000 miles per hour—which is fast enough to make the clocks on the satellites tick significantly slower than the clocks on Earth. The size of the effect slows the clocks in orbit relative to the ground by 7 microseconds per day: 350 times the precision needed to find a location to a meter. The good news is that the size of the effect is known with great precision from the equations for special relativity and can be corrected for. This is built into the software used by the receivers to obtain the best accuracy.

CAUSALITY

One outcome of relativity was a new understanding of causality. Most of us have a sense of causality in our minds—if you do X, then Y will happen. We also have embedded in our psyches the notion of time being a one-way street that we cannot travel in freely, unlike space. In special relativity, space and time get mixed together in a very specific way that makes the speed of light an ultimate speed limit that no physical object can either exceed or attain. This also sets a limit to the influence of events on each other. If lightning strikes, the flash takes a finite amount of time to reach our eyes. We are only aware of the strike when the light reaches our eyes; otherwise we're unaware of it.

In special relativity two events can only influence each other if they exist at places and times where the separation allows light to communicate between them. Otherwise, there is no influence. That is to say, there is no such thing as faster-than-light communication or influence. It takes eight minutes and twenty seconds for light to travel from the Sun to the Earth. The nearest star to us is Proxima Centauri, which is about four light-years away. If there is alien life on a planet circling the star, and they try to send us a signal, it takes four years to reach us. Before that, there is no link.

Figure 8.14 illustrates this. Imagine a pebble dropped into a pond at some point and circular ripples emanate. Like the ripples, if an atom emits light, spherical light waves move through space from the atom. Two events can be causally linked if they occur inside this moving light sphere of influence. There is also an implication of the one-way arrow of time that only something in the past can influence something in the future. Although there are novels about time travel, it seems like we are trapped within the one-way street of past-and-future causality. The horizontal axis in figure 8.14 represents time, and the vertical axis represents space. The place at the intersection of the axes is a moment in time. It can only influence things in the future if it lies inside its *future light cone*. Only events inside the *past light cone* can influence that particular moment.

In the case where two events exist inside the light cone, it is called a *time-like* separation, meaning that one can influence the other. When two events exist outside the light cone, they cannot influence each other, and this is called a *space-like* separation.

It's difficult to wrap one's mind around space-like separations, but imagine that at this moment aliens start to send radio waves toward Earth from a planet circling Proxima Centauri. We can't possibly know of this at this moment or up to four years into the

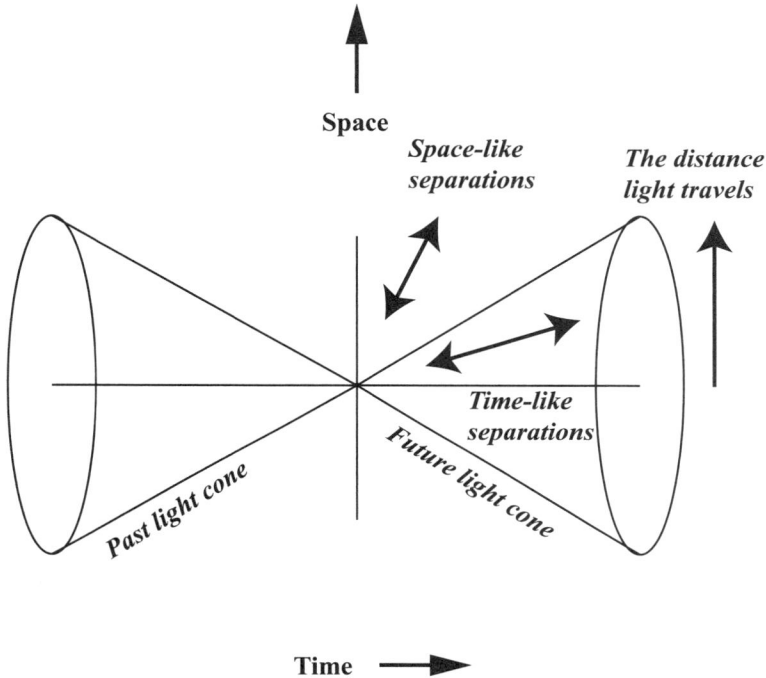

FIGURE 8.14 The light cone of causality.

future, because it takes that long for the signals to travel to us. Our moment *now* on Earth, and the start of the radio transmission from Proxima Centauri are two space-like separated events. On the other hand, the release of a baseball from a pitcher's hand and the contact with a bat are time-like separated events because light can connect the two events.

You can only influence things in your future, not your past. This is *causality*. What's odd about physics is that the formulation of our laws does not make this distinction, at least at the moment. One example is light. If we switch on a lightbulb, the light waves move outward. This is one of two possible solutions in the equations of physics. There is another solution where the light rays move inward to the lightbulb. Since this isn't observed, we eliminate that backward solution "by hand," and move on to other things. The arrow of time remains a mystery. Movie maker Christopher Nolan wrote and directed a movie, *Tenet*, where he imagined the ability of people to willfully travel backwards in time and the possible ramifications.

A concept about how an object can move through space-time is a *worldline* (or world line). We think of objects of having an identity: a person, the Earth, an airplane, even if they consist of a collection of components like atoms. As an object moves through space-time, we can think of the motion as a series of causally connected events that trace out a path in both space and time. The main constraint is that the object's worldline must be within its light cone (fig. 8.15).

One example is the Earth's orbit around the Sun. From a spatial perspective, it describes a big circle around the Sun, ending up

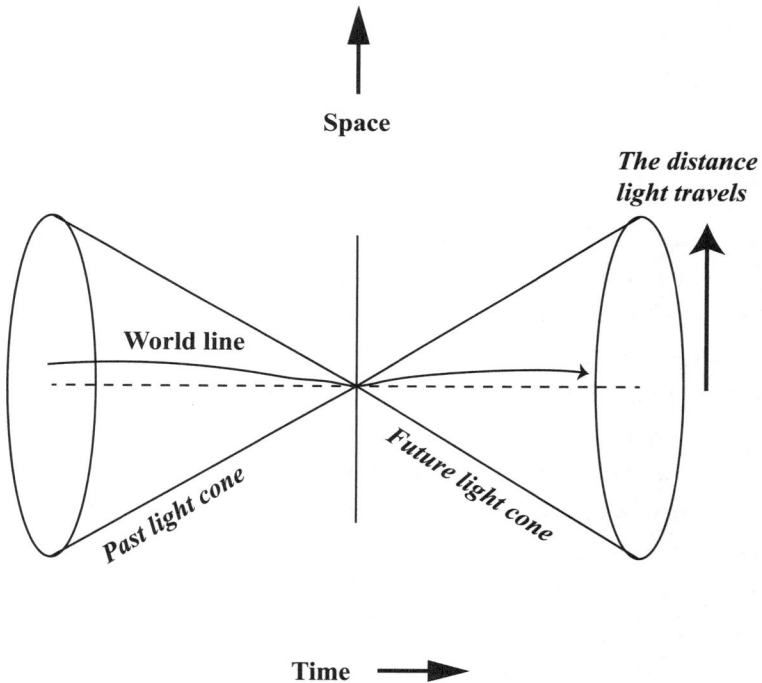

FIGURE 8.15 An object tracing out a worldline in space-time. Objects can be thought of a series of causally connected events.

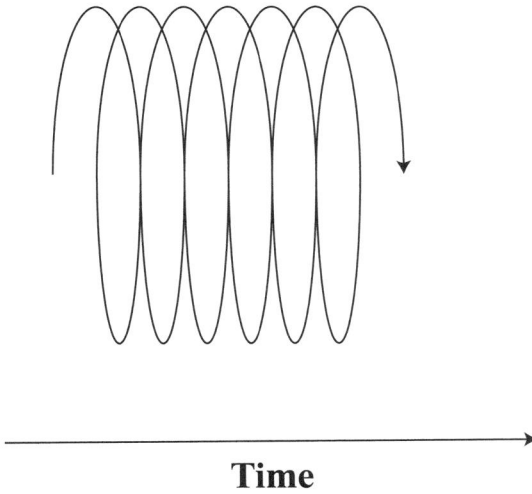

Time

FIGURE 8.16 Earth's orbit through space-time has a worldline path that
looks like a helix. This is from the frame of reference of the Sun.

in the same place once a year, but if we add time as a coordinate it
has a pathway that looks like a helix as it moves through space-time
(fig. 8.16). This is from the frame of reference of the Sun, and we're
assuming that the Sun is at rest, even though we know it's orbiting
the center of the galaxy.

SPECIAL RELATIVITY AND CULTURE

Einstein published his paper on special relativity in 1905. As word
spread, relativity became extremely popular. Lectures to lay audiences by Einstein were in high demand. He wrote a popular treatise,
Relativity, The Special and General Theory, for a public hungry to
understand what these new theories meant.[8]

Einstein became an immediately recognizable cultural icon with
his unruly hair and mustache. His musings on things far afield from
physics were captured in memorable quotes, and he used his celebrity
to advance noble causes.

Probably one of the most memorable part of relativity is the famous equation:

$$E = mc^2$$

Just as the concepts of distance and time get modified when objects move close to the speed of light, concepts of energy and momentum likewise had to be modified. When we reformulate these to be consistent with special relativity, the masses of particles must get factored in. This equation gives a recipe for how mass can get converted into energy and became synonymous with a new era of all things nuclear: energy, weapons, and propulsion systems.

One particular fallout of the popularity of Einstein and the theory of relativity was the sense that everything is relative. As physicists we often try to explain difficult concepts in human terms, using terms like "observer" and "clock." Einstein's popular rendition builds his discussion on an observer on a train and one on the ground, similar to my development of Sam and Jess and their twins above. These kinds of descriptions, intended to convey the concept to others, paved the way for the idea that truth is relative for different observers, and there is thus no absolute truth.

Although the idea that truth is relative and not absolute has been debated for quite some time, the notion experienced a rebirth with relativity in the first half of the twentieth century. This was not Einstein's intent, but the cat was out of the bag. Below is one example of how the meaning of the special theory changed as popular explanations were interpreted and reinterpreted:

Author Peggy Rosenthal, in her book *Words and Values*, writes about special relativity, quoting mathematician Martin Gardner's rendition of the topic intended for general audiences. In many of Einstein's lectures and popular writings he used two flashes of lightning as observed in two frames to motivate the question of simultaneity, just like I invoked the idea of two firecrackers above. Rosenthal, quoting Gardner, who is explaining Einstein, writes:

The question of whether the flashes are simultaneous cannot be answered in any absolute way. The answer depends on the choice of a frame of reference. . . . This is not just a question of being unable to learn the truth of the matter. There is no actual truth of the matter. . . . There is no question of one set of measurements being "true," another set "false." Each is true relative to the observer making the measurements; relative to his frame of reference.[9]

Rosenthal then interprets Gardner further herself:

Seen in this form, "relative's" main act as the star of Einstein's theory seems to be to strike a quick karate chop at the solid Newtonian concepts of absolute length and absolute time, splitting the "absolute" right off from them. Hurling absolute into the abyss of "no meaning," "relative" then picks up the pieces of "length" and "time" that are left and attaches them to itself.

The emphatic denial of "actual truth" is the climactic epistemological moment of relativity.[10]

I liken the above to a version of the "game of telephone," where a phrase gets garbled when being passed on from among many individuals. The starting point would be Einstein's original paper, *Zur Elektrodynamik bewegter Körper* (On the Electrodynamics of Moving Bodies), which was translated from German. Then, there is Einstein's popular treatise, "Relativity, the Special and General Theory." Martin Gardner interprets Einstein's rendition for lay readers. Finally, Rosenthal takes Gardner's account and draws her own conclusions. In each step of the process, something changes until "actual truth" is denied. Just to be clear, Einstein never suggested that his relativity somehow negates "actual truth." In fact, he expressed some horror at the misuse of his relativity.

In the first half of the twentieth century, disciplines bearing the names "relativism" and "relativity" emerged: cultural relativism and linguistic relativity to name two. It's not clear if either of these was directly linked to special relativity, but it's possible that the *zeitgeist* of the era contributed to the rise of many "relatives."

Historian Paul Johnson, in a retrospective of the modern era from 1920 to 1990, saw a direct linkage between relativity and a broader kind of relativism:

> It was as though the spinning globe had been taken off its axis and cast adrift in a universe which no longer conformed to accustomed standards of measurement. At the beginning of the 1920s the belief began to circulate, for the first time at a popular level, that there were no longer any absolutes: of time and space, of good and evil, of knowledge, above all of value. Mistakenly but perhaps inevitably, relativity became confused with relativism.[11]

One relativism, cultural relativism, evolved quite independently from special relativity. This was championed by anthropologist Franz Boas, who advised that people engaged in ethnographic studies necessarily brought the prejudices of their own culture into the studies of others. Along with fellow anthropologist Bronisław Malinowski, Boas developed a methodology of the anthropologist immersing himself or herself in the culture being studied: living with the people and speaking their language, as a way of shedding the biases inherent in coming from a different culture.

Another kind of relativism is *linguistic relativity*, often called the Sapir-Whorf hypothesis after Edward Sapir and Benjamin Lee Whorf. The idea posits that language influences thought. The way language is constructed forces people to think along the rules embedded in grammar. Sapir was a student of Boas, and Whorf later expanded on Sapir's ideas. Whorfianism, as it is sometimes called, has created no shortage of academic debates on language, its universality, or lack thereof, and whether it is deterministic of thought. One example goes back to chapter 1, where there is the possibility that being forced to articulate directions in terms of east, west, north, and south in a language makes the speakers better at navigation.

There is no obvious intellectual connection that links special relativity with cultural relativism and linguistic relativity, other than some inspiration from the philosophy of the eighteenth and nineteenth centuries. However, the use of the term "relative" and the concurrent rise of these ideas in the 1920s may have conspired to create a sense that everything was relative.

Einstein was appalled that his work would be misunderstood and invoked in odd contexts. By his own account, and as discussed, he was influenced by philosopher David Hume. In turn, philosophers attempted to invoke relativity. H. Wildon Carr, a professor of philosophy at Oxford invoked special relativity to attack a broad philosophical position called "materialism." As a summary, materialism is the notion that matter is the only "real" thing in the universe. Matter makes stars, it makes thoughts, and everything emerges from it. Carr claims that special relativity slays materialism:

> It may not be obvious at once that the mere rejection of the Newtonian concept of absolute space and time and the substitution of Einstein's space-time is the death-knell of materialism, but reflection will show that it must be so. . . . For the concept of relative space-time systems, the existence of mind is essential. To use the language of philosophy, mind is an a priori condition of the possibility of space-time systems; without it they not only lose meaning, but also lack any basis of existence. . . . The concrete unit of scientific reality is not an indivisible particle adversely occupying space and unchanging throughout time, but a system of reference the active centre of which is an observer co-ordinating his universe.[12]

This seems to arise from a natural misunderstanding. Einstein's construction, particularly the popular rendition, invoked observers looking at clocks. The observers are not necessary for the transformation laws but invoking them in the explanation helps one's understanding. In some way the necessity of an observer ties back to an egocentric perspective. Perhaps it goes without saying, but the physics of the early universe did not contain conscious observers at every point in space and time, yet special relativistic effects were most certainly present. Carr's is an ironic position, considering that Einstein was influenced by Mach, who himself rejected dualism.

Another philosophical viewpoint, called "perspectivism," was considered to be validated through special relativity by the Spanish philosopher and essayist Jose Ortega y Gasset. In effect, perspectivism rejects absolute truth and maintains that truths depend on one's perspective. Gasset writes of the shift from Galilean relativity into special relativity:

> The theory of Einstein has shown modern science, with its exemplary discipline—the nuova scienza of Galileo, the proud physical philosophy of the West—to have been laboring under an acute form of provincialism.[13]

The relativism that emerged in the 1920s persisted well through the end of the twentieth century, with various incarnations such as postmodernism and critical theory. Many of these viewpoints centered around the concept that knowledge was related to the cultural context in which it was embedded. Although much of the advent of relativism was more "in the air" than directly inspired by special relativity, there was curious return to it in the 1990s in the "science wars." When postmodernists began dissecting the practices of scientists through their own critical lens, scientists began to push back.

Famously, special relativity was invoked multiple times in the context of postmodernism. One particular quote by philosopher Jacques Derrida became something of an epistemic football during the science wars:

> The Einsteinian constant is not a constant, is not a center. It is the very concept of variability—it is, finally, the concept of the game. In other words, it is not the concept of something—of a center starting from which an observer could master the field—but the very concept of the game which, after all, I was trying to elaborate.[14]

The "Einsteinian constant" is the speed of light. Critics of critical theory said that this quote demonstrated a clear lack of knowledge of physics on Derrida's part, while defenders said that the critics lacked knowledge of Derrida, taking it out of context.

With the idea that all truth is relative, it raises the question of "relative to what?" If truth depends on the culture in which it's expressed, as per the postmodernists, could there be an even finer-grained relativity of truth?

In an article in *The Atlantic*, "How America Lost Its Mind," Kurt Anderson reports about the trend in the 1980s and '90s toward cultural relativism in American academia:

> Relativism became entrenched in academia—tenured, you could say. Michel Foucault's rival Jean Baudrillard became a celebrity among American intellectuals by declaring that rationalism was a tool of oppressors that no longer worked as a way of understanding the world, pointless and doomed. In other words, as he wrote in 1986, "the secret of theory"—this whole intellectual realm now called itself simply "theory"—"is that truth does not exist."[15]

With the rise of social media from about 2007 onward, any individual could make truth claims to the public at large. With the emergence of global communities of "friends" and "followers" who can self-select, new aggregations of groups were possible. These new groups on social media could cluster around certain topics of common interest and, not surprisingly, political leanings. Accompanying this was the rise of online "news" sources that simply invented stories that they felt would attract readers and, in turn, advertising revenue. The secret to the success of these online sites was their ability to spin stories that appealed to certain segments of viewers who would then pass these on to their online clusters. The only validity necessary was an "echo-chamber" effect where constant repetition was sufficient to guarantee a sense of truth.

By the time of the US presidential election in 2016, this tendency to generate false statements online and propagate them was elevated to a new high, and any sense of truth online became highly subjective. A new word emerged to describe the era, "post-truth." "Post-truth," as an adjective, is defined by the OED as:

> Relating to or denoting circumstances in which objective facts are less influential in shaping public opinion than appeals to emotion and personal belief.

It is not entirely clear what dynamics led to the post-truth era. Certainly, the rise of social media was one factor. Another factor may have been the "echo-chamber" effect from repetition. Some writers blamed postmodernism for the rise of the post-truth era. It's not entirely clear that the intellectual circles of postmodernism and the online post-truthers had much overlap, nonetheless the superficial similarities in the attitudes toward "truth" led to conjecture of causation linking them. As an example, here is a snippet from a piece written by writer David Ernst, titled "Donald Trump Is the First President to Turn Postmodernism Against Itself," appearing in *The Federalist*:

> Many have argued that Trump is the product of political correctness (PC). This is true only in part. Rather, both PC and Trump's response to it are fruits of the postmodernism that has long ascended to the heights of our culture: the nihilism in the common presumption that all truth is relative, morality is subjective, and therefore all of our individually preferred "narratives" that give our lives meaning are equally true and worthy of validation.[16]

A BBC piece on philosopher A. C. Grayling went further:

> Appropriately for a philosopher, he [Grayling] identifies postmodernism and relativism as the intellectual roots "lurking in the background" of post-truth.

"Everything is relative. Stories are being made up all the time—there is no such thing as the truth. You can see how that has filtered its way indirectly into post-truth."

He says this has unintentionally "opened the door" to a type of politics untroubled by evidence.[17]

With such a curious historical arc, it is easy to lose sight of the fact that the aim of special relativity was to develop a transformation law that kept the speed of light a constant for all observers.

GENERAL RELATIVITY

As we saw earlier, Einstein was influenced by Mach. One of the important influences on Einstein's theory of gravity is what we've come to call *Mach's principle*. In it Mach ponders the influence of all the mass of the universe. You're staring at the night sky, standing still with your arms hanging by your side. You see stars all around and then start to spin relative to the stars. As you spin, your arms rise from what we might call a centrifugal force. But the rising up of the arms is in relation to all the stars in the sky, the universe at large, somehow defining a preferred frame of reference with respect to acceleration.[18]

Special relativity is the case of frames of reference moving with relative, but uniform velocity. What about the case of accelerated frames of reference? There are two fundamental pieces of information in Newton's laws that Einstein reexamined:

1. The "a body in motion tends to remain in motion"—in a more formal way of speaking, this means that with no force acting on an object, it will move with a constant velocity. That is to say, in a straight line with no change in speed.
2. "Force equals mass times acceleration." The mass in this statement is the resistance to acceleration, given some force. This is called the *inertial mass*.

But this is the same mass that appears in the gravitational force law, and we call it the *gravitational mass*. We might think that the two have nothing to do with each other. Because the inertial and gravitational mass are identical, the two masses cancel each other out, which is why objects fall at the same speed regardless of their mass. This isn't necessarily the case. Take electromagnetism. The force on a charge is independent of mass and is a different characteristic of matter. The cancellation wouldn't take place.

Gravity is unique in that the attractive force is proportional to mass. Why inertial and gravitational mass are equal is a mystery, but we can take it as a given. There is a thought experiment involving a person in an elevator and standing on a scale (fig. 8.17). If the person is in a stopped elevator in a gravitational field, they will feel a downward pull. If we transport the person into outer space, where there is no gravitational field, but accelerate the elevator, the person will feel a fictitious force that is indistinguishable from gravity. This is much the same as the feeling one has going around a curve in a car, where the acceleration associated with turning creates the sensation that

Elevator stopped
In gravitational field

Elevator accelerating up
No gravity

Elevator in free fall
In gravitational field

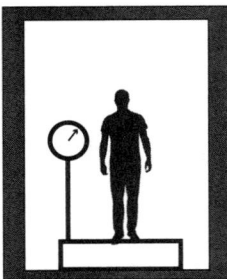

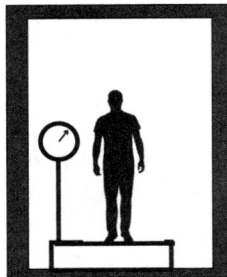

FIGURE 8.17 A person in a stopped elevator in a gravitational field will show some weight against a scale. If the person is transported into free space and accelerated at the right rate, they will also register a weight on the scale. If the person is put in the elevator in the presence of gravity and allowed to fall freely, they will appear to be weightless.

a force has been switched on. Finally, if a person is in an elevator that is freely falling in a gravitational field, it is as if gravity has been switched off.

This kind of thought experiment led Einstein to believe that an accelerating frame of reference was indistinguishable from a gravitational field—known as the *equivalence principle*. When an object is not experiencing a gravitational force, in free fall it is said to be following a *geodesic* path. Normally, in the absence of any force, this geodesic path is just a straight line with a uniform speed, but in the presence of a gravitational field the path is curved.

If we return to the trajectory of a cannonball in flight, we know that this is a parabola. If we were launched in an elevator in a gravitational field on this free-fall parabola, we would be weightless. In fact, this is the way that jet planes are used to simulate weightlessness for astronauts in training—they fly in parabolic arcs (the plane is known as the "vomit comet," for obvious reasons).

In general, the trajectories of objects feeling a force are curved. In a gravitational field objects in free fall don't seem to experience a force and are said to be following their geodesic paths. In Einstein's theory of relativity a gravitational field is equivalent to a curvature of space-time.

This curvature is general, so that light will also follow geodesics. Again, in the elevator analogy, a light beam going into an elevator that seems to be accelerating will appear to be bent. Likewise, from the equivalence principle, light would bend in a gravitational field, following a geodesic. Figure 8.18 shows a thought experiment about the beam of a flashlight shining through a hole in an elevator accelerating upward, but also stationary in a gravitational field.

So what creates the gravitational field in the first place? Mass, just like in Newton's laws. In general, the larger the mass, the greater the curvature of space-time (fig. 8.19). When I say "space-time" there is a natural implication that time is also affected, and it is—clocks run slower in a high gravitational field than they do in space that's free from gravity.

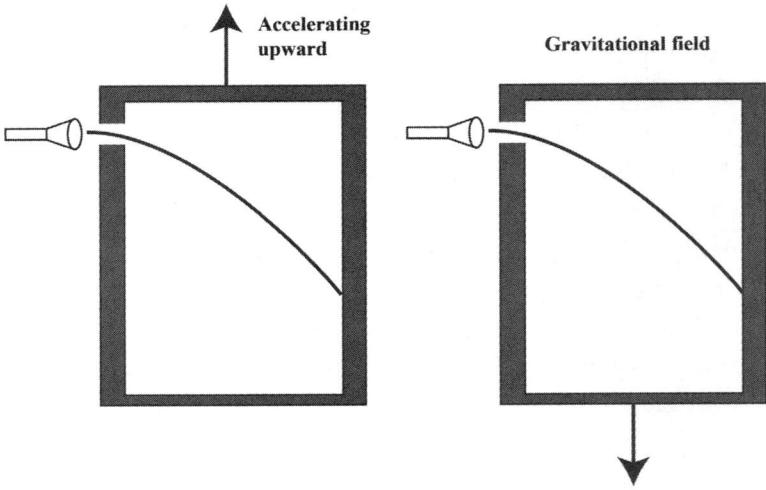

FIGURE 8.18 A prediction of general relativity is that light will appear to bend when seen in an accelerating frame of reference and will likewise bend under the influence of gravity.

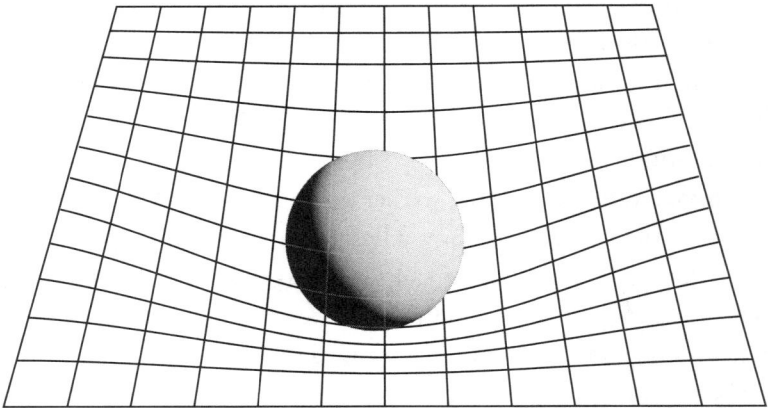

FIGURE 8.19 Bending of geodesic lines under the influence of a mass is akin to a marble placed on a rubber sheet and distorting it.

The theory of general relativity is developed by employing the language of curved surfaces, but now that we're talking about space-time, rather than space alone, time gets caught up in the formulation.

There are many predictions of general relativity that have been borne out—here's an incomplete list:

- Bending of starlight around the sun. This was one of the earliest tests of general relativity and was verified during an eclipse of the sun in 1919.
- Precession of the perihelion of Mercury. This was more like a post-diction, since it was already known by Einstein when he was formulating relativity–but the calculations are spot on.
- Gravitational radiation. When electric charges are accelerated, they emit light. A prediction of general relativity is that when masses are accelerated, they emit gravitational radiation—a twitch in the fabric of space-time that moves at the speed of light. Gravitational waves have been observed and are now being recorded for insights into astrophysical processes.
- Time dilation in a high gravitational field (i.e., clocks tick more slowly).
- Gravitational lensing, where the images of distant galaxies get distorted as the light from them passes through the curved space of intervening galaxies.

There are many more predictions of general relativity, too many to be listed here, but every test of general relativity has been borne out so far.

General relativity predicts even more exotic objects. One is a *black hole*. Here the gravitational force is so strong that not even light can escape. Another is a *wormhole* where two distant parts of space can be linked together by a kind of shortcut. As discussed, science fiction has invoked wormholes as a kind of interstellar transport that can evade the vast distances in outer space.

In the limit of small masses, general relativity becomes exactly Newtonian gravity. This means that general relativity doesn't "replace" Newtonian gravity, just that it has Newtonian gravity as a low-mass limiting case. Also, in the limit of small masses, we end up with special relativity. Finally, for low masses and relative speeds much less than the speed of light, we end up with Newton's physics and a flat space and absolute time. Newton's physics and gravity remain as a limiting case of general relativity.

At some level, we've come full circle since the time of Newton and Galileo. While the space in that era was somehow independent of the objects occupying the space, space now becomes defined by the objects occupying that space. A colleague of mine remarked that this is a bit like the difference between soccer, which is played on a fixed field, and water polo, where the playing field moves with the players.

There is an even more profound shift in our views of space from general relativity. What Einstein did not first appreciate but was embedded in his equations was the possibility of descriptions of universes. This is not just one kind of universe, but many kinds of universes. The structure of the universe is defined by the matter and energy contained in it.

This is somewhat odd that space at first was defined by the Earth at the center and the spheres of the planets surrounding them. Then we graduated to a space with the Sun at the center and the planets orbiting them. Then we had a kind of agnostic space that existed independent of matter. And now we have a space that's defined by the matter and energy inhabiting it.

* 9 *

THE STAR RECKONING

In the 1920s a fortunate convergence of relativity and observations gave us our modern model of the universe. When Einstein initially proposed general relativity, he was describing local effects, like the bending of starlight passing near the Sun. On the other hand, his equations allow the possibility of a description of an entire universe, even different varieties of universes. Perhaps one of the more daunting aspects of our current view of the cosmos is wrapping our brains around the mind-blowing size and age of the universe. It almost defies comprehension.

THOUGHTS ON THE UNIVERSE: *THE SAND RECKONER* AND OLBERS'S PARADOX

Before diving into a contemporary understanding of the universe, let's first take a look at some earlier concepts on the imagined scale. Around 200 BC Archimedes wrote *The Sand Reckoner*, where he pondered an upper limit to the number of grains of sand that would fit into the universe. This work contains one of the mentions of the model of Aristarchus with the Sun at the center of the solar system (quoted in chap. 2).

The Sand Reckoner seems remarkably modern in its construction. To make his estimate of the number of sand grains, Archimedes devised a system of large numbers reminiscent of our use of exponents of 10. He used the number *myriad* (μυριάς) for 10,000 as a basis. To get to larger numbers, he writes of a "myriad of a myriad," which would be $10,000 \times 10,000$ and so on.

Archimedes first had to estimate the size of a grain of sand. Then the difficult part was an estimate of the size of the universe. He assumed that the universe is finite, and that the outermost boundary was a sphere of fixed stars. The distance to the fixed stars was unknown, and the Earth–Sun distance was estimated by Aristarchus in relation to the Moon–Earth distance.

To make progress Archimedes had to make an important assumption about the distance to the sphere of fixed stars: that the ratio of Earth's diameter to the Earth's orbit was the same as the ratio of Earth's orbit to the sphere containing the fixed stars (fig. 9.1). Even in Archimedes's time reasonable estimates of Earth's circumference could be made. In the end Archimedes concluded that the universe could contain no more than 10^{63} grains of sand. In modern units Archimedes estimated that the fixed sphere of stars was roughly one light-year away.[1]

While Archimedes suggested the universe was finite but large, others like the atomist Lucretius imagined it might be infinite in extent. What would that imply?

The possibility of an infinite universe led to unusual speculations. Although he was not the first, in 1823 astronomer Heinrich Wilhelm Olbers articulated a paradox that bears his name. Olbers imagined what it would be like if we inhabited a static infinite universe with an infinite number of stars in the sky. Extending a ray out from the Earth into space, the ray would eventually terminate on a star. If this were the case, then the entire sky, day or night, would be as bright as the surface of the Sun (fig. 9.2). Since the night sky is dark, something about Olbers's paradox must be wrong. The poet Edgar Allan Poe, in his essay *Eureka*, wrote about it in 1848:

Were the succession of stars endless, then the background of the sky would present us a uniform luminosity, like that displayed by the Galaxy—since there could be absolutely no point, in all that background, at which would not exist a star. The only mode, therefore, in which, under such a state of affairs, we could comprehend the voids which our telescopes find in innumerable directions, would be by supposing the distance of the invisible background so immense that no ray from it has yet been able to reach us at all.

The "answer" to the paradox is that the universe is neither static nor infinite. Rather, it is expanding, and the more distant galaxies recede at higher and higher speeds. Ultimately, there is a cosmic

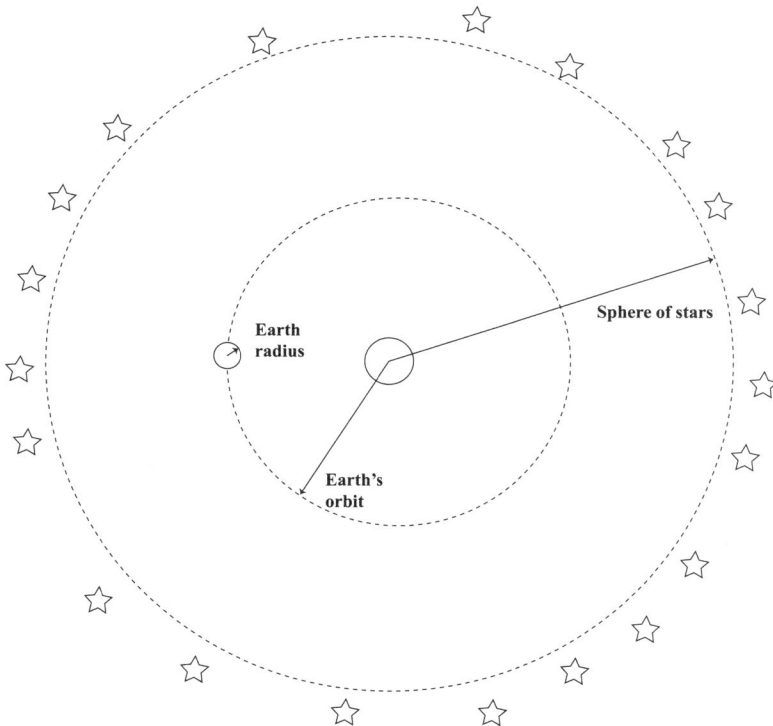

FIGURE 9.1 The universe from Archimedes's *Sand Reckoner*. The ratio of the Earth's radius to its orbital radius is equal to the ratio of the orbital radius to the radius of the sphere of fixed stars.

FIGURE 9.2 Olbers's paradox asserts that if the universe is static and infinite, eventually light rays from all stars would land on the eye and the night sky would be as bright as the Sun during the day.

horizon where distant galaxies recede at the speed of light, and we have no causal contact beyond this point. In other words, we never see the galaxies past this horizon.

THE COSMIC DISTANCE LADDER

To understand the large-scale structure of the universe, we need to know how to assess distances to objects like stars and galaxies. The *cosmic distance ladder* concept uses successive measures that overlap into larger and larger distance scales. We've already seen the first rung in the cosmic distance ladder in chapter 7 with parallax measuring the distance to the nearest stars.

With distances established using parallax, a kind of star called a *Cepheid variable* was employed to establish the next rung in the cosmic distance ladder. If you know the distance to a star, you can calibrate its absolute brightness, or luminosity. The apparent visible light from a star will fall off with distance in a well-understood relation, the same way a lightbulb will appear more and more dim the further away you go.

Cepheids are a kind of star that have a well-established brightness that varies with time, with periods ranging from a day to 100 days. Long-period Cepheids have higher luminosities. By observing the period of a Cepheid, and measuring its brightness, the distance can be established as the apparent brightness (observed brightness) falls with distance.

In 1890 Harvard established an observatory at Arequipa, Peru, to survey the part of the sky that was unreachable from the Northern Hemisphere. The surveys were done by imaging stars on glass photographic plates. Much of the analysis was performed at the Harvard College Observatory using "human computers," actually women, many of whom were recruited from Radcliffe College. Their job was to compile data based on scanning the images.

Henrietta Swan Leavitt was a computer, and she scanned many photographic images of the Small Magellanic Cloud, only visible from the Southern Hemisphere. She found a number of Cepheid variable stars in her scans and noticed a relationship between the variability period and the brightness. The Small Magellanic Cloud is at a large enough distance from Earth and compact enough to assume that all the stars are at the same distance from Earth, at least to a good approximation. This means that the relative brightness of the stars can be taken on face value. She then established the luminosity versus period curve for the Cepheid variables, which is a crucial rung in the cosmic distance ladder.

Knowing this, astronomers could then look for closer Cepheids that overlap with stars that have known distances from parallax measurements and establish an absolute distance scale that could be extended to far galaxies where Cepheids could be imaged. This increased substantially the radius from the Earth where the distance to galaxies could be measured: out to approximately 20 megaparsecs (Mpc) or 60 million light-years.

Beyond this, there is a kind of exploding star, a supernova, SN1a, that can be used to extend further the distance measurements beyond Cepheid variables in distant galaxies. White dwarves are stars that accrete mass from a companion star, causing them to grow in mass. There is a limit to how large white dwarves can grow, called the Chandrasekhar limit, approximately 1.4 times the mass of the Sun. At this limit, the burning of lighter elements into heavier elements becomes unstable and the star explodes. Since the stars all explode at roughly the same mass, their brightness curves are the same. They're much brighter than Cepheid variables and can extend the distance scale by about a factor of 500 into space beyond. This cosmic distance ladder of star parallax to Cepheid variables to supernovas allows us to measure the structure of the universe quite far out into space.

RECESSION OF GALAXIES

Over the course of the nineteenth century, telescopes became more powerful and different celestial objects were noted. One important feature was the appearance of nebulae ("mist" in Latin). While stars appear to be point-like celestial objects, nebulae are more spread out. There are different kinds of nebulae. Some consist of luminous gases spread out into space when a star explodes, such as the Crab Nebula. In the nineteenth century one type of nebula, with spiral shapes that look like a pinwheel or a whirlpool, were observed. Unlike the nebulae from exploding stars, which are relatively close by in our galaxy, the spiral nebulae are a great distance away, but this wasn't known until the 1920s.

In 1920 a debate began between astronomers Harlow Shapley and Heber Curtis, sometimes called "The Great Debate." Shapley maintained that the Milky Way was the entirety of the universe and that all the nebulae were glowing gaseous formations contained within or at least very close by. Curtis, on the other hand, argued that the Milky Way was just one of many of the spiral nebulae that are seen and that these all exist at a large distance from our galaxy. Figure 9.3 is a photo of the closest galaxy, M31 or Andromeda. On a clear night out in the country the Andromeda Galaxy can be seen as a faint fuzzy patch in the constellation Andromeda, next to the great square of Pegasus. With a pair of binoculars, you can make out some of the disk-like structure.

To resolve the Great Debate, astronomer Edwin Hubble trained a powerful telescope at the Mount Wilson Observatory on the

FIGURE 9.3 The Andromeda Galaxy. *Source*: NASA.

Andromeda Galaxy. There, he identified several Cepheids that allowed him to calculate the distance as 900,000 light-years, much farther than the size of the Milky Way, which is approximately 100,00 light-years across. This resolved the debate in favor of Curtis. Hubble sent a letter to Shapley describing his early results, and Shapley is reported to have said to a colleague, "Here is the letter that destroyed my universe."[2] Current measurements put the distance to Andromeda at more like 2.5 million light-years.

The term "galaxy" was originally applied to our Milky Way, and in fact the Greek root word *gala* means "milk." Spiral nebulae ultimately became termed "galaxies" when it was recognized that they were distant cousins of our own Milky Way.

THE UNIVERSE, ACCORDING TO EINSTEIN

When Einstein published his first findings on curved space-time, he made references to local effects, such as the bending of starlight in the presence of a gravitational field. But his equations also had interesting solutions: universes. One of the early solutions to Einstein's equations that described a universe was by physicist Alexander Friedmann, who in 1922 derived one that had an expanding space-time. The same solution was independently derived by physicist and priest Georges Lemaître and published in 1927.

In the 1920s Hubble continued with his studies of the distance of spiral nebulae using Cepheids. Many of these spiral nebulae were small and fainter, leading to speculation that the faint ones were farther away.

Another tool that emerged around this time was the use of spectroscopy to measure specific atomic lines. These spectra tell a curious tale of the relative motion of a distant galaxy to ours and employs the Doppler shift. Most of us experience the Doppler shift as the elevated pitch of a siren approaching us, with a lowered pitch as it moves away. For galaxies, atomic spectra can be used as markers of the motion. If the galaxy is moving toward us the color of its observable light spectra gets shifted toward the blue, while if the galaxy is moving away it gets shifted into the red.

The first measurement of a Doppler shift of a galaxy was done by astronomer V. M. Slipher in 1912, looking at the Andromeda Galaxy from the Lowell Observatory. He found that Andromeda was blue-shifted and was moving toward us at 190 miles per second. By 1925 forty-five galactic Doppler shifts had been cataloged. The list ranged from a blue shift of 190 miles per second (blue-shifted) to 1,125 miles per second (red-shifted).[3]

Data from Cepheids allowed Hubble to correlate the Doppler shifts with distances to the spiral nebulae. On average, the more distant a galaxy was, the more red-shifted it was. There was a considerable range of both blue and red shifting but the net result was consistent with a larger recessional velocity the farther a galaxy was from us.

It seems strange that our galaxy would not be in motion and the farther away you look the faster galaxies seem to be moving away. This would seem to imply that we are at the center of the universe—or at least in a privileged position. But this doesn't make a lot of sense. This is where the *cosmological principle* comes into play. The concept is that if you were transported to a distant galaxy, you would see roughly the same thing—your galaxy would appear to be "at rest," and the more distant a galaxy was, the faster it would go. This also is sometimes referred to as the *Copernican principle*, which more broadly means that we are not in some special place in the universe.

The cosmological principle makes for an interesting social metaphor that I call the human cosmological principle. In this metaphor, think of the social forces at play in your own life—it seems like you are singular, that your situation is unique. This is true to some extent. But now imagine that a huge number of people are experiencing life just as you are and it's only by virtue of your position of place that it seems to be unique. This is something of a metaphor that favors empathy: to imagine being transported into someone else and to look at the world through their eyes and realize that to each other person, the world also seems singular and unique from their perspective.

The relation between galactic distance and recessional velocity is what the Friedmann-Lemaître model predicted, and that the universe was indeed expanding. The rate of expansion became known as the Hubble constant, the ratio between the recessional velocity and the distance to a galaxy. Numerically, the Hubble constant is roughly 70 (km/sec)/Mpc (megaparsec).

Explanations of the expanding universe to a lay audience are often made with an analogy to an expanding balloon (fig. 9.4). Imagine a two-dimensional sphere/balloon where the surface is peppered with galaxies. You're only allowed to inhabit the surface of the balloon, knowing only two dimensions and not the third to which you're expanding. As the balloon expands, the galaxies on the surface recede from each other, and the farther galaxies recede faster than the nearer ones.

This balloon explanation works reasonably well insomuch as there is a spherical symmetry to our universe and objects on the surface of the balloon move away from each other the farther apart they are. The problem with the balloon analogy is that it misleads the

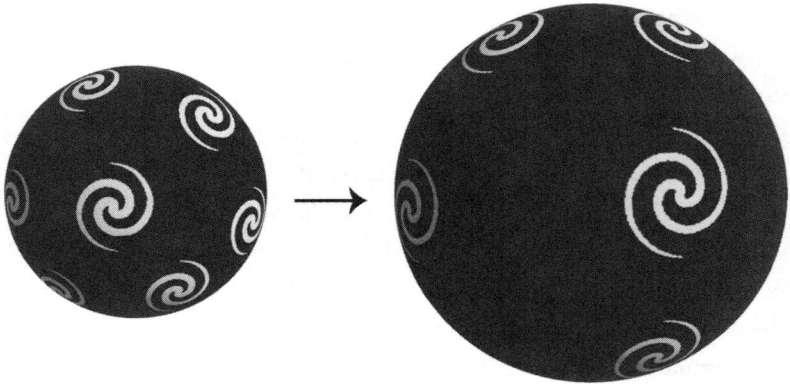

FIGURE 9.4 Depiction of the expansion of the universe
in the manner of an expanding balloon.

reader to believe that there is some additional dimension into which
the universe expands. This is the case for a balloon, which is a two-
dimensional surface expanding in three dimensions. We live in three
dimensions, and the balloon analogy might give a false impression
that there is a fourth dimension that we're expanding into, which isn't
the case. It is space itself that is expanding, and the distance between
galaxies that is growing.

Another analogy for this model that's frequently used is of a raisin
cake, with raisins embedded inside a dough. Over time, the dough
rises, increasing the distance between the raisins. This is perhaps a bit
of a better analogy, but the raisin cake is still expanding into something. In some sense, the cake itself is the domain of the universe.

There are some wrinkles on the description of our universe as
the Friedmann-Lemaître model. Their model is perfectly symmetric,
but our universe is somewhat lumpy, with local galaxies and galactic
clusters that make our universe an imperfect version of the model,
but on a large enough scale that it seems to work. By way of analogy,
the surface of the Earth has its own lumps in the form of mountain
ranges and valleys but when viewed from a large distance the Earth
is close to a perfect sphere.

The Friedmann-Lemaître model answers the question posed in Olbers's paradox: the universe is finite. Moreover, the recession of distant galaxies Doppler shifts their light further and further into the red. Now, if galaxies are moving faster and faster as we go farther and farther away, at some point the recessional velocity reaches the speed of light, at which point no observations are possible. By using the Hubble constant, a nominal radius can be calculated where galaxies recede at the speed of light. This is approximately 14 billion light-years. Care has to be taken in quoting this number because other factors come into play, but it gives a rough scale for the observable universe.

The volume associated with this distance is called the "Hubble volume" and is 10^{31} cubic light-years. We could return to *The Sand Reckoner*. Roughly estimating the volume of Archimedes's sphere of fixed stars as 1 cubic light-year, we could guess that we could fit 10^{94} grains of sand into the observable part of our universe.

There is a more physical concept known as the cosmological horizon. This is the maximum distance that light can travel to reach us. We believe that there was a very rapid expansion of the early universe that was so fast that patches of space were put out of contact with each other. Light cannot travel and connect these patches, and the farthest distance we could be in contact with is at this horizon.

There is another curious aspect of the Friedmann-Lemaître model. If we run time backward, all the matter in the universe becomes concentrated at a single point of infinite density. Running it forward from that moment, there would seem to be a vast explosion bringing our universe into existence, commonly known as the *Big Bang*.

A question that is often imagined about the Big Bang model is "what happened before the Big Bang?" We don't know, but the model itself has an answer. Recall in the last chapter there was the idea of a *worldline* of objects through time and space. If we trace the worldlines of particles backward in time, the Big Bang model suggests that all the worldlines converge at a single point. At that moment, there is no "before," only an "after."

At first blush this seems a bit crazy but consider the Earth and lines of longitude. If we were to travel toward the South Pole, just a few steps away from the South Pole we can identify north, south, east, and west, but precisely at the South Pole there is only one direction: north. One cannot speak of "south," at the South Pole in the same way one cannot speak of a "before" at the moment of the big bang, at least in the Friedmann-Lemaître model.

Is this point of infinite density real? We don't know, and it's quite possible that something different happened in the very early universe. It's worth bearing in mind that the Big Bang model is just that: a model.

The rate of expansion that we currently see is determined by several factors. First, there's the initial condition of the Big Bang, for example, how much energy is released. Second, there's the amount of matter in the universe. Depending on how much matter is in the universe, the expansion could slow down and the universe begin contracting, or the expansion could continue forever. Finally, there is the possibility of something called vacuum energy, which is an energy associated with free space. The vacuum energy provides a pressure that could accelerate the expansion of space. Soon after Einstein's initial publication of general relativity, he considered the possibilities of solutions that represent a universe. Thinking that the universe is static, he introduced a famous additional term, the cosmological constant, denoted by the symbol Λ (Greek Lambda). In Einstein's thinking the constant would keep the universe from contracting under the influence of gravitational attraction. The term is associated with vacuum energy and is now embedded in our standard cosmology.

COSMIC RADIATION AND BEYOND

In the 1960s astrophysicists contemplated more observational effects from the Big Bang. One was a residual electromagnetic radiation. The early universe consisted of an intense soup of free charged particles like electrons and protons. There also was a bath of photons, parcels of electromagnetic energy, that bounced between the charged particles. As the universe expanded it cooled, and the charged particles settled down to form neutral atoms after 380,000 years but there was still an intense amount of electromagnetic radiation left over that could now travel freely through space. As space expanded, the wavelengths of the photons associated with the radiation got stretched out, and their energies cooled. At the present epoch in cosmic expansion, the average temperature of the cosmic radiation is just a few degrees above absolute zero. This cosmic radiation was first detected in 1964, and since then progressively detailed measurements are providing significant information about the early universe.

There is one peculiarity about the cosmic radiation. This is not terribly meaningful but still a curiosity. Recall that Aristotle believed the Earth was motionless and the center of the universe. Newton's and then later Einstein's ideas were based on the notion that there is no preferred frame of reference that is at rest. However, there is a sense in which the cosmic radiation defines a kind of rest frame. The Earth is in motion with respect to the radiation. We can see this because there is a red shift of the cosmic photons in one direction in space and a blue shift in the opposite direction. The Earth is moving through this cloud of primordial radiation at a speed of 370 kilometers per second with respect to the cosmic radiation, which is a little over 10 times our orbital speed around the Sun. In principle, if we had a sufficiently powerful rocket we could launch it and put it in a frame that is at rest with respect to the cosmic radiation.

After the Earth's motion through the cosmic radiation is factored out, the temperature appears remarkably uniform—at approximately 1 part in 100,000. From a thermodynamic standpoint, this uniformity implies that the very early universe was in thermal contact. In other words, if one patch of the universe was hotter than another, the contact between the two would even out and place them at the same temperature in the same way a hot cup of coffee eventually cools down to room temperature.

This aspect is perhaps not remarkable to the casual reader, but there is a problem. Many parts of the universe are out of causal contact with each other. There is no way that a temperature from one patch could be communicated to another because they are separated by a distance that light cannot connect from one to the other.

One popular solution to this problem is called inflation. The concept is that the very early universe (about 10^{-33} seconds after the Big Bang) was small enough that it was in thermal equilibrium, meaning that light could readily move from one end to the other, creating a uniform temperature. Then a super-rapid expansion of space-time placed large chunks of the universe out of causal contact.

At this point, the casual reader may be wondering, "But I thought nothing could move faster than the speed of light!" In the theory of relativity, space can expand fast enough to create patches that are once in causal contact, but then fly away from each other so fast that light can no longer traverse the distance between them. It's matter that physically cannot exceed the speed of light.

The rapid expansion put large chunks of space out of causal contact with each other—which is how the separated patches would display such a uniform temperature. The end of the inflationary phase presents something of an issue. Inflation could just go on forever, but we're most definitely not in an inflationary phase. This problem goes by the name "the graceful exit problem." However the exit happens, as the different patches come out of inflation, it is believed that this creates the seed of structure formation.

An early model of inflation used a concept called quantum tunneling to explain its end and is now not favored but remains a possibility of how the universe could spontaneously change from one character to another. To explain tunneling, look at figure 9.5 with a surface that has two valleys separated by a mountain. The bottom of one valley is lower than another. Imagine that a ball is initially in the higher of the two valleys. The ball rolls back and forth about the bottom of the valley but doesn't have enough energy to get over the top of the mountain, which would allow it to drop into the lower valley. If we

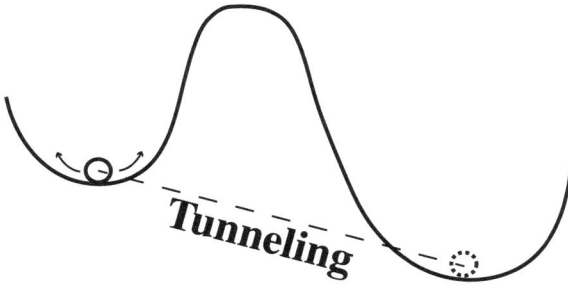

FIGURE 9.5 The process of quantum tunneling. In a landscape of energies there can be valleys and hills. Normally an object can be trapped in a higher valley and can't go over the hill into a lower valley, but if we add in quantum mechanics it can spontaneously "tunnel" into the lower valley. In the early universe the process of inflation is halted when parts of the universe tunnel out of the rapid expansion into a different phase.

add in quantum mechanics, there is a loophole, however. "Tunneling" would allow that ball to spontaneously appear in the lower valley without having to make it up to the top of the mountain and drop back down. The upper valley represents the conditions of the one phase and the lower valley represents another phase. There is some belief that our present universe could tunnel into another.

As we'll see in the next chapter, quantum mechanics has an element of randomness, and quantum tunneling is no exception.

THE DARK SIDE

One outstanding question in general relativity is what are the sources of gravity? In electromagnetism we speak of "source terms." Broadly speaking, these are the electric charges that give rise to electric and magnetic fields. In Einstein's relativity the source terms are both mass and a possible vacuum energy. To understand the structure and fate of the universe, we need to know the "budget" of the source terms. How much mass is there and how much vacuum energy is there?

Depending on the relative size of the cosmological constant to the amount of mass in the universe, the universe could collapse or expand, and the expansion could even accelerate. The acceleration is due to the vacuum energy being an intrinsic property of space. If you make more space by expanding, you make more vacuum energy, which drives more expansion, and so forth. We believe we are in a phase where a vacuum energy far weaker than that driving inflation appears and causes the expansion of the universe to accelerate.

In 1998 observations of the expansion of the universe pointed to an acceleration in the expansion. This was first seen in the recession of galaxies tagged by distant supernovas, but also appeared in measurements of the shape of the cosmic microwave radiation. We attribute this to a vacuum energy we have taken to calling *dark energy* or *quintessence*.

There are two perplexing questions that arise with dark energy—one is "what is it?" On this point, we don't know. The second question is related to the energy scale we associate with this mysterious energy. The dark energy we think we see is a huge number of orders of magnitude away from what physicists think would be a "natural" vacuum energy. The question is why nature has chosen such a small scale. This disconnect between what we believe in fundamental physics and the observations is unexplained. We also invoked vacuum energy to explain the inflationary period of the very early universe, but at a vastly different scale: much higher in energy, much shorter in duration.

I was on a panel, whimsically named "The Dark Energy Task Force," but nonetheless with a serious mission. Our mandate was to evaluate different possible experiments and observations to probe the nature and origin of dark energy. We made several recommendations, many of which are being carried out. Some astrophysicists have contemplated whether Einstein's equations are wrong, or in need of modification, but to date no viable alternative has emerged that is consistent with the array of astrophysical data.

Another mystery is something called *dark matter*. For gravity, there's the mass of ordinary matter, like protons, neutrons, and electrons. But there seems to be other sources of gravity out there. The first hint that there was something other than our ordinary matter came in 1933 from astronomer Fritz Zwicky. You can make predictions about the behavior of galaxies based on the assumption that the distribution of matter follows the luminous stars. Zwicky looked at a galactic cluster known as the Coma cluster and tried to calculate the velocities of the luminous matter assuming the visible galaxies in the cluster were the only thing that was there. He found velocities that were far in excess of the calculation. To account for that, he proposed that there was a huge amount of nonluminous matter, dark matter, that was spread out in a distribution that was much broader than the galaxies themselves.

The next hint that there was something other than normal luminous matter came from galactic rotation curves. If we assume that most of the matter in galaxies is distributed in the places indicated by stars, we can calculate that the matter's rotations rise rapidly as one moves away from the center and decline slowly as one move farther outward. From 1932 onward, however, more and more observations seemed to contradict this, and the rotational curves appeared to be much flatter after the initial rise. After making careful observations, astronomer Vera Rubin demonstrated in the mid- to late 1970s that the galactic rotation curves indicated a much broader distribution of matter than is clustered in the luminous central regions. Follow-up observations in many galactic systems confirmed the same behavior.

Yet another probe of dark matter is in gravitational lensing. This is when the image of a distant galaxy is distorted by the curvature of space caused by some intervening mass. By looking at the nature of the distortion of images, it's possible to make a map of the distribution of matter between the distant galaxies and the Earth. The lensing also shows broad distributions of dark matter between the more compact objects like visible galaxies. The details of fluctuations in the cosmic microwave radiation also point to the existence of dark matter.

Why is it referred to as "dark matter"? Mostly, because it doesn't follow the behavior of luminous matter that make up the visible stars.

What could it be? One candidate is a WIMP—which means "weakly interacting massive particle."

Why "weakly interacting"? If you accelerate a charged particle, it emits light. In galaxy formation there is initially a large amount of hot matter, including ions—charged particles. As the ions collide with each other they emit light and slowly cool down. The emission of light allows the charged matter to cool and coalesce under the influence of gravity. The very broad distribution of dark matter indicates that it doesn't interact like normal charged matter and must be more weakly interacting. One suggestion along the lines of a WIMP is a new form of matter that's charged, but has a very weak electric charge, perhaps one one-thousandth of the charge on an electron. Another possibility is a kind of particle known as a supersymmetric particle.

We've tried to produce dark matter in our highest-energy accelerators but have not yet found any evidence for it. If dark matter exists in the form of WIMPs, there is good reason to believe the Earth would be moving through a dark matter cloud, in the same way we know that the Earth has some motion with respect to cosmic radiation. If this is the case, a collision of a WIMP with an Earth-based atom would cause an observable recoil of a struck atom. Experiments deep underground with the purpose of finding WIMP have thus far come up empty.

Another dark matter candidate is a MACHO (yes, physicists have a strange sense of humor), meaning a "MAssive Compact Halo Object"—perhaps a halo of tiny black holes or tiny massive stars. These have partly been eliminated as a candidate in searches for "microlensing"—by looking for rapid changes in star luminosity when a MACHO passes between the star and a telescope on Earth.

Yet another hypothesis is MOND for MOdified Newtonian Dynamics, the idea that the laws of gravity (as we believe we know them) do not hold up at large distances. This has been proposed for the galactic rotation curves but doesn't seem to explain all the observations pointing toward dark matter.

For the current model we have of all the stuff in the universe, the source term is ΛCDM. The *CDM* part stands for "cold dark matter."

The Λ (Greek letter Lambda) is the symbol indicating vacuum energy. In the ΛCDM or Lambda Cold Dark Matter cosmology, our best estimates are that the universe is made up of approximately 70 percent dark energy, 25 percent dark matter, and 5 percent ordinary matter as we know it: protons, neutrons, and electrons (fig. 9.6). This cocktail of dark energy, dark matter, and ordinary matter fits well with many astrophysical observations. It's challenging that we don't know what 95 percent of the universe is.

It's tempting to speculate on the fate of the universe, and to wonder about the parts of the universe that we are currently out of contact with. The ultimate fate of the universe is tied to the amount of matter and energy contained in it. If there is a sufficiently large mass in the universe as a whole it will begin to contract in a scenario that's sometimes called the *Big Crunch*.

On the other hand, if the dark energy scenario is correct and if we run time forward, the expansion will continue to accelerate, and gradually more and more galaxies will be pushed out of contact with us. Locally, gravity is an attractive force, so it is unlikely that our little

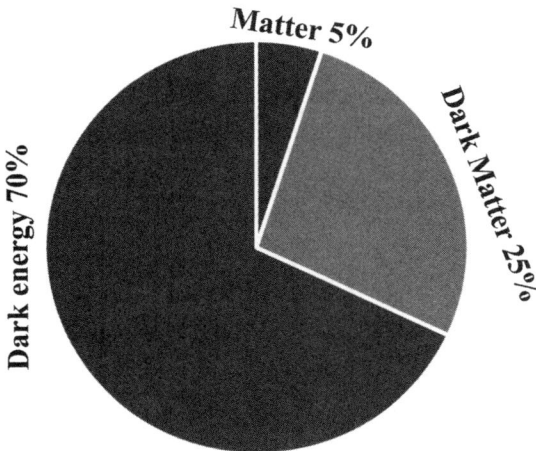

FIGURE 9.6 Matter and energy budget of the universe.

206 ★ CHAPTER NINE

patch of the universe in the Milky Way will disperse into nothingness, but we'll slowly become increasingly lonely as galaxies expand beyond the cosmological horizon.

Most of the ordinary matter in the universe is hydrogen (approx. 73%) and helium (25%). These both have their origins in the Big Bang. In the hot dense conditions in minutes after the start of the universe, protons and neutrons underwent nuclear reactions to form the isotope deuterium (one proton and one neutron) and the isotope helium-4 (two protons and two neutrons). On Earth, on the other hand, heavier elements like carbon and iron are, however, the byproducts of stellar evolution, including supernovas.

A major success of the Big Bang model is the explanation of the relative abundance of helium-4 to hydrogen in the universe at large. In addition, the ratio of photons to protons as about one billion is also predicted by the classic Big Bang model. Most of the physics of the reactions that give rise to the hydrogen/helium ratio and the photon/proton ratio are well understood and happened at a relatively well-defined phase of the early universe.

As of this writing, our ΛCDM seems to work reasonably well, but there are cracks appearing in the bulwark of modern cosmology. The value of the Hubble constant can be found from various observations. But when we compare these observations, they do not all converge on the same value. Rather, they disagree with each other at a level of significance that is becoming more and more difficult to ignore.

Another problem is found in recent observations of distant galaxies by the James Webb Space Telescope. Distant galaxies can be a probe for the early universe, and the telescope can image galaxies that had formed when the universe was only a few hundred million years old. The recent data indicate that the early galaxies are substantially heavier than our models of the early universe currently predict.

Do these tensions invalidate the ΛCDM model? Most aren't throwing our current model into the trash. There are many experimental uncertainties, and our modeling of the early universe probably isn't perfect. It speaks to the need for more efforts to see if we can resolve the tensions.

There are, however, some astrophysicists who suggest that these issues may be important enough to rethink our standard cosmological model. In an essay in the *New York Times*, physicists Adam Frank and Marcelo Gleiser suggest that perhaps it's time to reexamine the fundamental underpinnings of our modern cosmology.[4] They write:

> We may be at a point where we need a radical departure from the standard model, one that may even require us to change how we think of the elemental components of the universe, possibly even the nature of space and time.

For reference, figure 9.7 portrays a nominal history of the universe with significant landmarks or, perhaps more properly, spacemarks. After the universe comes out of the initial inflation, the first particles of normal matter show up after a microsecond. Most of the Big Bang nucleosynthesis occurs around the first three minutes. By 380,000 years, matter has coalesced into neutral atoms and cosmic radiation can now travel unimpeded. The first stars are thought to have formed roughly 200 million years after that, and then galaxies. We're currently in an era where dark energy is starting to dominate the budget.

THE ANTHROPIC PRINCIPLE

Vacuum energy is invoked for inflation at the very early universe. We don't know the mechanism for the vacuum energy, but if we assume it, it accounts for the structure of the universe as we observe

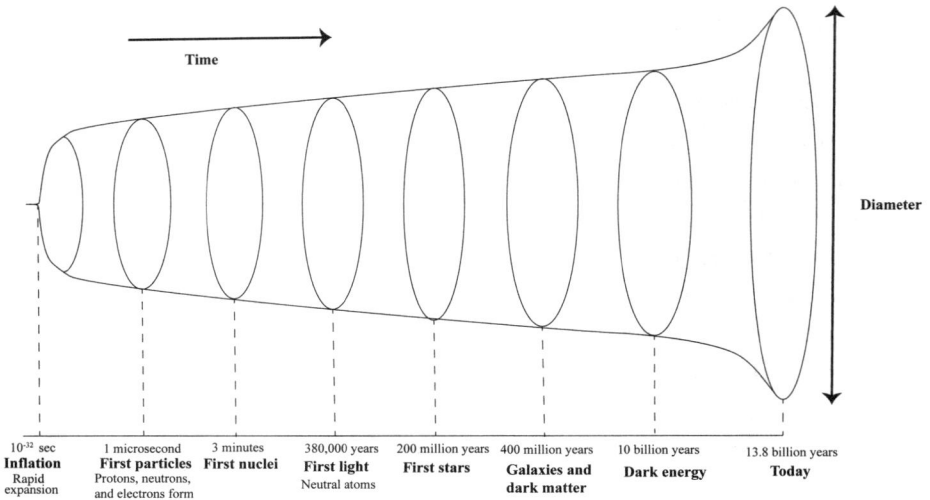

10^{-32} sec	1 microsecond	3 minutes	380,000 years	200 million years	400 million years	10 billion years	13.8 billion years
Inflation	**First particles**	**First nuclei**	**First light**	**First stars**	**Galaxies and dark matter**	**Dark energy**	**Today**
Rapid expansion	Protons, neutrons, and electrons form		Neutral atoms				

FIGURE 9.7 The understood history of the universe, with notable moments.

it. The vacuum energy drove that early rapid expansion. And, yet, we have what appears to be a very weak version of vacuum energy that is causing the acceleration of the expansion of the universe in our epoch some 14 billion years after the Big Bang. That we are even here is perplexing as the current vacuum energy is about 120 orders of magnitude weaker than what we might expect from particle theory. Said another way, the size, flatness, and age of the universe seems highly improbable given what we believe may have been in the physics at the time of the Big Bang. To get your mind wrapped around this improbability, imagine we filled up the entire visible universe with sand, à la Archimedes, and then asked for you to pick out one single particular grain of sand.

In fact, when we look at physical constants like strength of gravity, the speed of light, and the charge on the electron, our universe seems highly improbable, and yet here we are. A concept known as the *anthropic principle* has been invoked to explain what seems to be our unlikely existence. There is a *weak* and a *strong* form of the anthropic principle.

In the weak form the argument is that it is possible that many universes exist. In fact it almost seems mandatory that many universes exist. Each one has its own values for various physical constants. Since we are living in one universe, we just happen to be in one of the ones that possesses the capabilities of supporting intelligent life. In other words, our very existence only works out because we inhabit an amazingly lucky universe. In nearly all the other universes, there is no intelligent life because the proper conditions don't exist. By analogy we could look to our habitation in the *Goldilocks zone* in our orbit around the Sun. We exist on Earth because we're on a planet that happens to be in a not-too-hot, not-too-cold zone, and here we are.

The strong form of the anthropic principle is more egocentric in its nature than the weak form. The strong form posits that, in order to exist, not only must a universe possess the capability of forming intelligent life, but it must also create intelligent life. What this mechanism might be, we don't know and seems to verge on the religious. There is an even stronger form that posits that our existence in an improbable universe is proof that there is a divine creator, who crafted the fine-tuning in such a way that intelligent life will emerge.

What do most physicists and astrophysicists think about the anthropic principle? For some it's viewed as something of a cop-out. Rather than saying that there are things we don't understand yet, most believe we need to concentrate on the physical processes, like what is really in the mass budget of the universe, and the nature of fundamental physics. Living with uncertainty can be challenging. More on this in a later chapter.

The strong form of the anthropic principle and the notion of a divine creator can be appealing, for obvious reasons. It puts intelligent life in a position of primacy. But, on the other hand, many physicists would subscribe to the notion that we're just a speck in the void, and that's that. Theoretical physicist Steven Weinberg held that "speck in the void" view. In his book on the early universe, *The First Three Minutes*, he concluded, "The more the universe seems comprehensible, the more it also seems pointless."[5]

We'll return to the anthropic principle later, but we must first depart from the realms of the largest to the realms of the smallest. Already we've encountered gravity and electromagnetism, which are long-distance forces, but there are two other forces: the strong and weak forces. While electromagnetism and gravity seem to have an infinite reach, the strong and weak are confined to extremely short distances, on the scale of the atomic nucleus. And these two forces are absolutely crucial to our universe. Without the strong and weak forces, stars would not burn. We'll examine three of the four fundamental forces in a bit, but first we need to look at the rules that govern matter on the shortest distance scales: the quantum realm.

* 10 *

INTO *the* REALM *of the* SMALL: QUANTUM MECHANICS

There are curious parallels between Einstein's relativity and the development of quantum mechanics. Both emerged in the first half of the twentieth century. Both originated in considerations of the properties of light. Both describe domains that are beyond our everyday experience. On this last point, it also becomes challenging for a lay audience to understand the two theories. Like relativity, quantum mechanics has spawned some amusing social metaphors that physicists sometimes employ as an inside joke. On the other hand, serious misconceptions about humans and quantum mechanics are visited on the public at large by some who try to popularize the concepts but often go off track.

While general relativity describes the universe in the spatial domain of the very large, quantum mechanics is mostly in the realm of the tiny: on the scale of atoms, and nuclei. The condensed story of quantum mechanics is that it describes matter as having both a wave-like property and a point-like property. This is often referred to as the *wave-particle* duality.

Even as early as the seventeenth century, light was treated as a particle in some cases, and a wave in others. Light could travel from one place to another in what seemed like straight lines, perhaps getting bent by a prism or lens, but in predictable paths. But there were other cases, such as light passing through a narrow slit when the result looks like the waves from a pebble dropped into a pond.

In the late nineteenth century, we came to realize that light was an electromagnetic wave moving through space, as stated in the chapter on relativity. At the time the theory of electromagnetic waves developed, there were advances on questions of heat transfer: thermodynamics. The marriage of the two branches of physics landed on a major problem associated with something called *black-body radiation.* In thermodynamics we often talk about a system that is in equilibrium, namely that the temperatures are the same for objects in contact with each other, like a hot cup of coffee cooling down to room temperature over time. If you pour that cup of coffee and look at it with an infrared camera, it will initially glow reddish, while the rest of the room looks cooler. Then as the coffee cup radiates heat, it will eventually cool down and will appear to be at the same temperature as its surroundings. In the case of a black body, there is an object surrounded by electromagnetic radiation, like light or microwaves. The object and the radiation are in contact with each other and eventually come to an equilibrium, like the coffee cup and its environment. A characteristic shape of the distribution of temperature for the body and the surrounding radiation can be derived from assumptions of thermodynamics and the characteristics of light.

A major problem with the classical treatment of black-body radiation was that it yielded infinite results, while it was known from experiment that the spectrum of the radiation is finite. Physicist Max Planck tried fitting black-body data with different analytic forms. Inspired by the thought that perhaps the radiation came in discrete parcels he called quanta, he added a term to the fit and found that it gave a good account of the black-body spectrum. Ultimately, this solved the problem of the infinite results and gave a hint that light came in packets later called photons. Planck's fit also yielded a constant, h, that bears his name and gives the scale where quantum effects become important.

Planck's black-body radiation formula has been used countless times to assess the average temperature of bodies, such as the Earth or Sun. The cosmic radiation I described in the previous chapter follows Planck's black-body radiation formula. So, in this sense the entire universe is a black body with an average temperature a few degrees above absolute zero.

Einstein published his special theory of relativity in 1905. That year he also published a remarkable paper on an experimental result called the photoelectric effect. In experiments light of different frequencies was shined on metals, kicking out electrons. The energies of the electrons were independent on the intensity of the light but did change energy depending on the frequency. The higher the frequency, the higher the energy of the electron. This result was surprising at the time, but Einstein suggested that the quantum of light had an energy that was just Planck's constant times the frequency of that quantum.

At this point, the idea of a quantum of light was realized: both a particle *and* a wave. This means it can move in straight lines and is a bundle of energy with its energy being directly related to its frequency. By the 1920s this concept was extended to other particles, like the electron, and in fact all particles. The name *photon* appeared for a quantum of light during this period.

WAVE PACKETS AND THE UNCERTAINTY PRINCIPLE

If the readers have followed the discussion up to this point, they might be wondering: how can a thing be both a particle and a wave at the same time? The word *particle* conjures up the image of a point-like object, like a period at the end of this sentence. The word *wave* on the other hand, suggests something much larger. In fact, if there is a wave with a single wavelength or frequency, it conceivably could have an infinite spatial length associated with it.

One way out of this seeming paradox is the concept of a *wave packet*. This is a combination of waves of different wavelengths added together to create a wave-like object that inhabits a finite portion of space.

Figure 10.1 shows one example of a wave packet. As you can see, it has a finite spatial extent, and yet has a wave-like quality to it. In order to make such a packet, we need to add together a bunch of waves of varying wavelengths and sizes.

The addition of many constituent waves is shown in figure 10.2 where they add together to create the wave packet. The packet can only interact over its distance scale, which is called a *coherence length*. It is similar to the sound of a single note being played on an instrument, like a piano. You hear the onset of the note, the note for a brief period of time, and then it goes away.

The wave packet in figures 10.1 and 10.2 is one example of a wave function, which is a general term for quantum states. Another example of a wave function is the cloud of electrons bound to the nucleus of an atom.

We can use the wave-packet construction in figures 10.1 and 10.2 to describe a central tenet to quantum mechanics: the Heisenberg

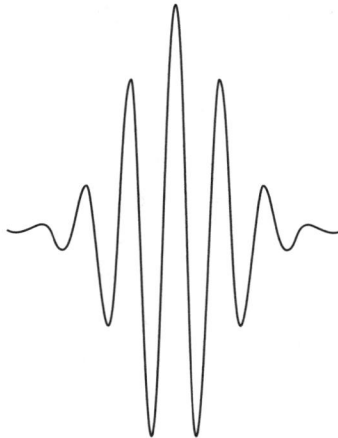

FIGURE 10.1 A wave packet.

Coherence length

Wave packet

Constituent Waves

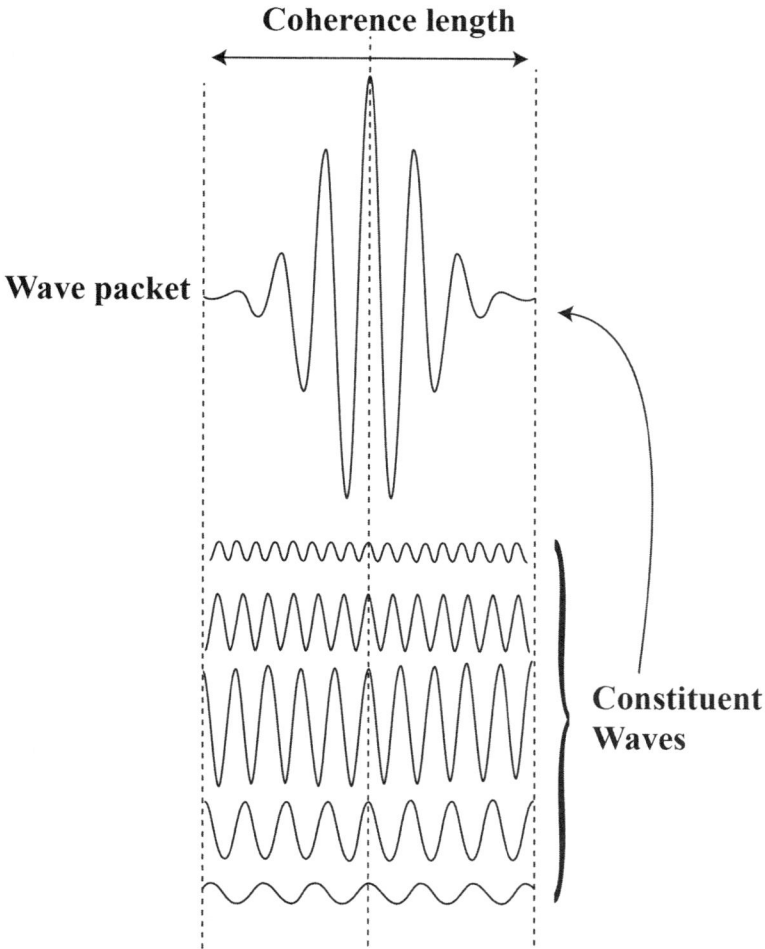

FIGURE 10.2 Illustration of how constituent waves add together to give a wave packet. Waves of different frequencies and different amplitudes can add together to create a wave packet of a finite coherence length. The constituent waves themselves have an infinite spatial extent.

uncertainty principle, named after physicist Werner Heisenberg. If we had only one wavelength contributing to the wave packet, the packet itself would have an infinite spatial extent (coherence length). On the other hand, if we wanted to create a wave packet that was precisely localized in a single spot in space, we would have to add together a very large number of constituent waves. There is a tradeoff between the localization in wavelength and the localization in space that is the result of the way wave packets get constructed.

The more localized a particle is in space, the less localized it is in frequency. As I noted above, the energy of a quantum of light is directly related to its frequency, so in a sense there is a tradeoff in the precision of nailing down a particle in space and the precision of finding a specific energy.

The uncertainty principle, as described above, is perhaps subtle until you think about how to create wave packets from component waves, and how they get localized. Nonetheless the very idea of an intrinsic "uncertainty," by virtue of the word, has created misconceptions, which have embedded the idea into popular culture. In the TV series *Breaking Bad*, a high school chemistry teacher named Walter White turns into a callous drug dealer who uses the code name Heisenberg. In an evident reference to the uncertainty principle, the name implies a moral fluidity, where a respectable man turns into a violent criminal lacking predictability.

One common misunderstanding of the uncertainty principle often appears in popular literature. The idea is that somehow the act of measuring influences the measurement itself. This definitely can happen and is often called the *observer effect*, but this is not the uncertainty principle, which, as described above, is the result of wave mechanics. Nonetheless this measurement idea is widely invoked. In the movie *The Lost World: Jurassic Park*, mathematician Ian Malcolm invokes this misunderstanding in a bit of dialogue with paleontologist Sarah Harding, when she objects to him lighting up a cigarette:

SARAH HARDING: Don't light that! Dinosaurs pick up scents from miles away. We're here to observe and document, not interact.
DR. IAN MALCOLM: Which is a scientific impossibility. The Heisenberg Uncertainty Principle. What you study, you change.

This snippet of dialogue has Dr. Malcolm confusing the uncertainty principle with the observer effect.

In an essay on misunderstandings of the uncertainty principle, philosopher Craig Callender writes about a New Age movie invoking the uncertainty principle to justify some dubious assertions,

> The film "What the Bleep Do We Know!?" uses it [the uncertainty principle] to justify many articles of faith in New Age philosophy. Asserting that observing water molecules changes their molecular structure, the film reasons that since we are 90 percent water, physics therefore tells us that we can fundamentally change our nature via mental energy.[1]

The uncertainty principle has been invoked in other social contexts, which often make little sense, as the original meaning gets rapidly confused. In an article, "Uncertainty About the Uncertainty Principle," Jim Holt writes: "No scientific idea from the last century is more fetishized, abused and misunderstood—by the vulgar and the learned alike—than Heisenberg's uncertainty principle."[2]

In *Physics World*, Robert Crease wrote about popular misunderstandings of the uncertainty principle:

> Browse through any bookshop's "new-age" section and you'll find peculiar statements about the uncertainty principle, including the claims that its implications are psychedelic and that is heralds cultural revolution.[3]

Crease goes on to quote a conversation between theater director Anne Bogart and famous vocal coach Kristin Linklater in the magazine *American Theater*,

LINKLATER: "Some thinker has said that the greatest spiritual level is insecurity."
BOGART: "Heisenberg proved that. Mathematically."
LINKLATER: "There you are."

MEASUREMENTS AND COLLAPSE OF THE WAVE PACKET

In the above I wrote about waves, but the question naturally arises—"a wave of what?" This brings to mind the ether that was believed to carry light. Einstein showed that this ether was not necessary if we changed the way we thought of space and time, particularly from the point of view of making measurements or observations. In a similar way, for quantum mechanics the operational act of making a measurement gives some insight into the meaning of the wave.

Let's think of the wave packet of light being one of many photons from an object impinging on the sensors of a digital camera. The camera sensors can measure the position of the pixel and the energy deposited in that pixel. The energy will give a frequency and, hence, color of the light. But . . . you can only light up one pixel at a time (fig. 10.3). Typically, photons that we deal with on Earth, say, illuminated by sunlight or from a lightbulb, have coherence lengths on a human scale, like the size of your hand. The camera sensors, on the other hand, can be very small, much smaller than the coherence length of the photon.

So, before detection, the photon is spread in space, and the sensors on the camera are blank, ready for an interaction. Then, the photon

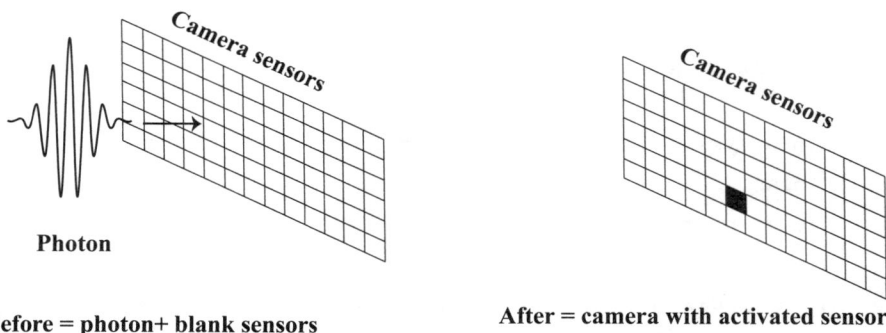

Before = photon+ blank sensors   **After = camera with activated sensor**

FIGURE 10.3 The measurement process using a photon and camera sensors as an example. The photon is much larger than the size of the camera sensors but appears on one sensor at a time.

interacts and is detected on one of the sensors. The photon has disappeared, and what is left is the position of the sensor and a record of the energy of the photon. What happened to the photon?

The "recipe" for interpreting the wave function is that when you square it, you get values that are probabilities. With one interaction only, you sample part of that distribution of probabilities. The photon is spread in space, but you are sampling one of those possible positions. Likewise, the photon is made up of a range of frequencies (energies), and you're sampling one of those. Over time, with repeated measurements of identical photons, you slowly build up a picture of the wave function. You sample all possible positions and all possible frequencies (energies).

In a sense the wave function is a wave of a probability of existing, which may or may not be a satisfying statement. The moment when the photon disappears and a sensor shows a result is called wave-packet collapse. The idea that the square of the amplitude gives a probability of a measurement is usually attributed to physicist Max Born. It's a simple and instrumental way of comparing calculations to measurements. This seeming randomness of the process bothered Albert Einstein, who famously mused that "God does not throw dice."[4]

The equation describing quantum mechanics, including wave packets and the wave functions of atoms, was developed by physicist Erwin Schrödinger in 1925. This provided the mathematical backbone of calculations but shed no light on the notion of how the wave-packet collapse might occur. Max Born's statistical interpretation is considered an important axiom of quantum mechanics that allows us to connect the wave-packet construction with measurements.

The probability associated with the wave function has been something of a nerdy inside joke about where we spend our time, at least among physicists. This metaphor holds up rather well. Once I ran into a colleague from the United States at the restaurant at the European Center for Nuclear Physics (CERN) where I often travel for my experiment. I asked how he was doing, and he said "Okay, but

I feel like my wave function peaks over the Atlantic." This was his way of saying that he was traveling a huge amount between his home institute and CERN. You can extend this into your own life, perhaps about spending time at work, at home, shopping, your favorite vacation spot. Of course, you'd have to have someone in on the joke and they'd have to know what you mean by a "wave function," which is probably a stretch.

OBSERVERS

During the development of quantum mechanics, a particular school of philosophy was quite popular in Europe, called Logical Positivism, championed by a group of philosophers known as the Vienna Circle. In this branch of philosophy, only empirically verifiable or observable phenomena were considered "real" in some sense. Many of the physicists developing quantum mechanics felt compelled to always make contact between the theory and actual observables—even to the point of elevating the importance of the observer to a lofty role. In this line of reasoning, a photon may interact with an atom, but the spread of the wave packet is preserved in the interaction, and the uncertainty information gets passed on to a related uncertainty in the atom. This continues until an observer intrudes, and then the wave packet collapses the entire chain. The logic is that consciousness, however defined, sits beyond the realm of quantum mechanics and plays a special role.

It is important to pause here and make a few points about this view:

1. At no place does the mathematics in quantum mechanics lead to the need for an observer for a wave packet to collapse.
2. At the time of the development of quantum mechanics, there was not a clear definition of what it means to be "conscious." In fact, psychologists and neuroscientists still struggle with the concept.

A number of conjectures exist, and they're worth exploring, but there is nothing definitive.

3. The concept of a conscious observer is something of an add-on. There's no mathematical account of what it means—only words spoken or on pages.

On the possible origins of this notion: the implications of quantum theory were discussed at the First Copenhagen Conference (Easter 1929) where physicist Charles Galton Darwin initiated the viewpoint on consciousness. Here is a summary of what Darwin said, from the recollections of physicist Leon Rosenfeld:

> He argued that the seat of an observation could be shifted from the physical receptor to the retina, and thence to somewhere in the brain, where we are absolutely compelled to stop. Right up to the brain the process is non-committal: it is only after our consciousness has animated the proceedings that it is possible to infer back and describe what actually happened in the familiar language of particles. Thus, we have a sub-world described by a wave-function, a dead world, not involving definite events, but instead the potentiality for all possible events. It becomes animated by our consciousness, which so to speak cuts sections of it when it makes observations.[5]

Although Rosenfeld found the concept questionable at the time, this idea of consciousness being required for a quantum mechanical measurement was slowly baked into early interpretations of the new theory. In an early treatise on quantum mechanics (1932), physicist John von Neumann mentioned this idea about consciousness and referred to a "cut" between the physical world and the consciousness of the observer as something he called a psycho-physical parallelism, and maintained that it is a crucial underpinning of science. This causation between a large-scale event like conscious awareness and wave-packet collapse has sometimes been associated with the

Copenhagen Interpretation—both after Dane Niels Bohr, and the 1929 conference where it had its origin. The Copenhagen Interpretation is best described really as a cluster of interpretations describing quantum mechanics and a grappling with the probabilistic nature of the wave-packet collapse, but it is sometimes associated with the effect of consciousness.

Some of the early ideas about possible linkages between consciousness and quantum mechanics became so set into our history that it is difficult to disentangle this viewpoint, even though there are variants of wave-packet collapse that do not invoke consciousness to validate quantum measurement.

In a New Age book, *The Tao of Physics*, author Fritjof Capra wrote about connections between Eastern mysticism and modern physics. On the role of the observer, he says,

> The human observer constitutes the final link in the chain of observational processes, and the properties of any atomic object can only be understood in terms of the object's interaction with the observer. This means that the classical ideal of an objective description of nature is no longer valid. The Cartesian partition between the I and the world, between the observer and the observed, cannot be made when dealing with atomic matter. In atomic physics, we can never speak about nature without, at the same time, speaking about ourselves.[6]

The very language of quantum mechanics contains relics of the possible conjoining of the quantum world with conscious observers. When we talk about measurements, physicists refer to "observables," which already conjures the image of an observer.

SCHRÖDINGER'S CAT

It isn't axiomatic that physicists make good philosophers, or vice versa. Erwin Schrödinger, who developed the equation describing wave functions, struggled with the question of wave-packet collapse and interpretations that invoke the role of a conscious observer. In 1935 he proposed a famous thought experiment involving a cat in a box with a sinister device.

A cat is put into a sealed box, with a radioactive source, a Geiger counter, which amplifies charged particle energies, and a poison. For the time the cat is in the box, there is a 50–50 chance that one of the atoms in the source will decay. If it does decay, the decay product, a charged particle, will enter the Geiger counter, releasing a cascade of electrons, which will trigger an electric circuit that will release the poison, killing the cat. Schrödinger then muses whether an observer opening the box will kill the cat by observing it—the idea being that inside the box there is a superposition of a 50 percent alive cat and a 50 percent dead cat. The intent of the thought experiment was, in part, to demonstrate that there was a ridiculous limit to extending the notion of wave-packet collapse all the way up from atoms to the scales of human observers. But, surprisingly, over the years, people thought the opposite of the famous Schrödinger's cat, and believed that the observation kills the cat, and it has become part of popular culture.

One resolution of the issue raised by Schrödinger has to do with the typical coherence lengths in the box, the timescale of the observation, and whether the entire box can be thought of as a single quantum entity. Photons from a candle might have a coherence around, say, 10 centimeters. Some lasers can produce photons with coherence lengths of kilometers. Atoms, on the other hand, can be quite localized with a size of 10^{-10} meters, although they can be made to interact on much larger scales.

In more recent formulations of quantum mechanics, there is the idea of decoherence: when a quantum system interacts with its environment, the wave function loses coherence over time, and interference can no longer take place. As to the cat in the box, the coherence length (and time) of the nucleus that produces the decay that kills the cat is quite short, and as soon as the Geiger counter receives the decay product from the nucleus, an irreversible amplification occurs where a cascade of electrons emerges that releases the hammer. All the atoms in the hammer, the vial, the cat, and the box have coherence lengths at the scale of a nanometer and can't possibly produce interference at the dimensions and timescale of the box and the observer. In other words, either the nucleus decays or it doesn't, and the cat dies or it doesn't—but opening the box has nothing to do with its death.

In the example of the photon and the camera screen (fig 10.3), on the other hand, the coherence length of the photon is the same scale as the camera, and the interaction with the sensors on the camera screen is responsible for the collapse of the wave packet. The time between the emission of the photon and its detection is about a billionth of a second, which is quite different from the relevant timescale of the cat in the box. Quantum coherence is maintained in the case of the photon and camera. We can imagine how physics might have evolved if photons of light didn't possess such long coherence lengths. This is somewhat akin to wondering whether a theory of gravity might emerge in a solar system with three stars in chaotic motion, as described in the novel *The Three-Body Problem* that I referred to in chapter 7.

Although from time to time the question of consciousness intrudes to establish the collapse of the wave packet, practicing physicists rarely give it much thought. As physicist David Mermin related, "If I were forced to sum up in one sentence what the Copenhagen Interpretation says to me, it would be 'Shut up and calculate!'"[7]

To this day, connections between quantum mechanics and consciousness are still articulated. It is sometimes invoked in discussions of free will versus determinism. The probabilistic nature of wave-packet collapse provides a nice metaphor for decision-making, but that doesn't imply that our decision-making is driven by quantum mechanical effects.

Returning to Schrödinger's cat, it is perhaps one of the most popular cultural manifestations of quantum mechanics. In part it conjures the divide of making decisions—that, prior to a decision, there is a pregnant moment when future paths appear before us. It has appeared as plot devices in the cinema, theater, science fiction, even poetry.

Poet Peggy Landsman writes in "Schrödinger's Cat":

Though I'm only a thought in his mind,
It is taught I'm a curious kind.
Not here and not there,
I pop up everywhere
Demonstrating one cat's double bind.[8]

A popular wrinkle on Schrödinger's cat is the phrase "Schröding-er's douchebag" or "Schrödinger's asshole." I only learned this recently. This is a case where someone (the "douchebag") ridicules or makes fun or slights another person. Depending on whether the other person objects or not, the "douchebag" will say, "I'm only joking, can't you take a joke?" but only if there is an objection.

QUANTUM ENTANGLEMENT

Another popular social metaphor arises from a phenomenon known as *quantum entanglement*. In a quantum system with multiple parti-cles, the wave functions can be said to be *entangled* if the wave func-tion of two or more particles are interdependent. If there are two entangled particles, a measurement of one particle collapses the wave functions of the other.

Here's an example. Imagine an atom emits two photons simul-taneously that travel in opposite directions. The wave packets of the photons will be entangled with each other—in particular their polarizations are linked. Figure 10.4 shows the possible polarization states of a photon. A photon is an electromagnetic field that is oscillating and there are two possible polarizations—in the figure one vibration is vertical and one is horizontal, but both are per-pendicular to the direction of motion. The photons are linked so that if one is measured with a horizontal polarization, the other must be vertical and vice versa, but each is a 50–50 mixture of the two states.

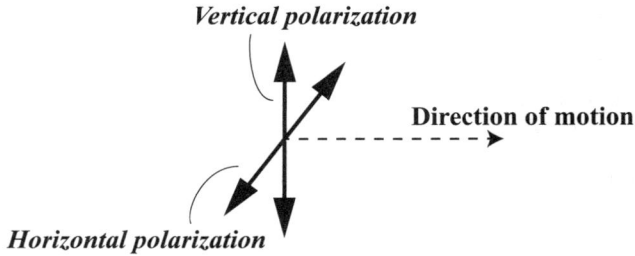

FIGURE 10.4 Two possible polarization states of a photon.

Figure 10.5 shows the ways the measurements can turn out. There is a 50 percent chance of measuring the photon in one of the two polarization states in one of the two detectors placed away from the atom that emits them. As soon as one photon is measured, the wave packet of both photons collapse. The photons could be very far away from each other, but there is still a linkage of the two. In fact, the collapse of one wave function brought on by the measurement of the other has to happen faster than the speed of light. How does one "know" about the other? This seems to violate common sense, and Einstein referred to this as "spooky action at a distance." But there is a sort of special dispensation for entangled objects, just like there's a kind of dispensation for patches of the universe retreating faster than the speed of light. Effectively, there is no transmission of information from one detector to the other.

The concept of quantum entanglement has bred, and not doubt will continue to breed, social metaphors. One version is love. Two lovers are entangled in such a way that if one is distant and experiences some pain or dies, the other instantly knows about it without receiving any news. Closely related is the concept of soul-mates. Often one sees cases where life partners die in close succession with one another. This is an entanglement of sorts.

The metaphor can work both ways. In popular explanations of quantum entanglement, it has been referred to as "love on a sub-atomic scale."[9]

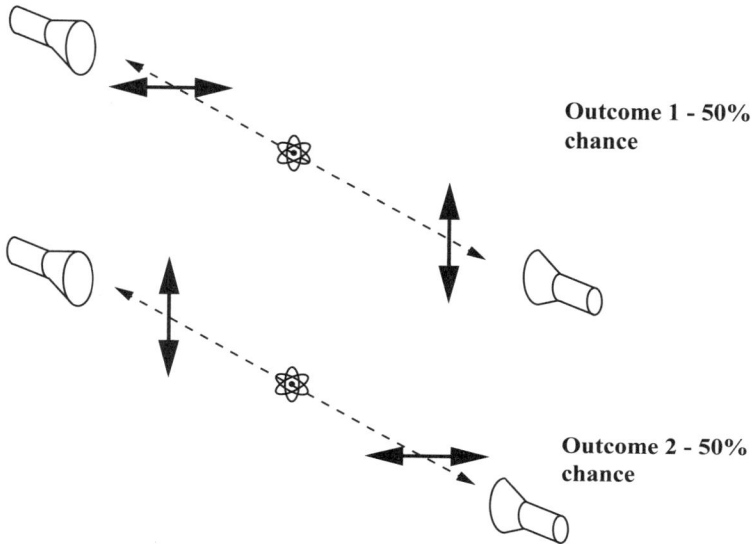

FIGURE 10.5 Quantum entanglement in two photons emitted from an atom. They have a 50-50 chance of being polarized either horizontally or vertically, but there is a correlation. When one is polarized horizontally, the other must be polarized vertically.

Language has been invoked in a human-scale entanglement. In "Quantum Mind and Social Science," Alexander Wendt explores this. In an interview Wendt explained, "The interesting part of this for me is the non-separable idea, the entanglement notion. The idea there is that even though our bodies are encased in skins and are separate from each other, our minds are entangled through language, and that language, I argue, is a quantum phenomenon that entangles our minds together."[10]

In an article, "Humanism, Science, and the Radical Expansion of the Possible: Why We Shouldn't Let Neuroscience Banish Mystery from Human Life," writer Marilynne Robinson laments the decline of the humanities as inspiration. She writes, "The spirit of the times is one of joyless urgency, many of us preparing ourselves and our children to be means to inscrutable ends that are utterly not our own. In such an environment, the humanities do seem to have little place."

She continues, and finds solace in quantum entanglement,

> The last century and the beginning of this one have without question transformed the understanding of Being itself. "Understanding" is not quite the right word, since this mysterious old category, Being, fundamental to all experience past, present, and to come, is by no means understood. However, the terms in which understanding may, at the moment, be attempted have changed radically, and this in itself is potent information. The phenomenon called quantum entanglement, relatively old as theory and thoroughly demonstrated as fact, raises fundamental questions about time and space, and therefore about causality. . . . However pervasive and robust entanglement is or is not, it implies a cosmos that unfolds or emerges on principles that bear scant analogy to the universe of common sense.[11]

While it is interesting to take quantum entanglement as a metaphor for human thought and behavior, it's quite another thing to posit true quantum interference on our scale of being. In the movie *What the Bleep Do We Know*, a literal human entanglement is asserted with something called a "universal wave function." Our thoughts and intentions supposedly interact with this wave function. If we think good thoughts, it affects the universe positively; if we think bad thoughts, it affects the universe negatively. The movie advocates a constant state of positivity to influence the universe. Positive thoughts and quantum entanglement have also been proposed as a path to good health by some advocates. This line of reasoning has come under strong criticism from physicists as quackery.

A discussion of the social metaphors arising from quantum mechanics would not be complete without a mention of the existence of numerous self-help books that invoke the topic as a kind of insider's knowledge. These are too numerous to get into, but I'll mention one by way of example. Author Laura Berman has advocated quantum mechanics to create a satisfying love life. Here is a clip from an advertisement for her book *Quantum Love: Use You Body's Atomic Energy to Create the Relationship You Desire*,

Now love, sex, and relationship expert Laura Berman, Ph.D., taps the latest scientific and metaphysical research to offer an inspiring alternative: a higher level of love beckoning you to move forward, not backward. Using the essential truth we've learned from the study of quantum physics—the fact that at our molecular core, each of us is simply a vessel of energy.[12]

Not surprisingly, pushback against these notions has extended beyond skeptical scientists. Political activist Barbara Ehrenreich in a lecture, "Smile or Die," took strong exception to the idea of a perpetual state of positivity, calling it a delusion:

> The delusional extreme embedded in all this is the idea that you change the physical world with your thoughts. When you send thoughts out from your mind they exert a force that brings things to you that you want. So, if I were doing this right we could all concentrate on getting a million dollars or whatever and indeed it would come. And there have been a lot of attempts to explain this scientifically. . . . Now they talk about quantum physics. I love that in [popular renditions of] quantum physics for some reason it's become an excuse to mock all of science. See if nothing's real, nothing's true and whatever you think, that's how the world is. So, if you think positively, you remake the world positively according to this pseudo-scientific explanation. But, anyway what's wrong with this? Why not delude yourself into thinking everything's fine and that you can change the world with your thoughts?[13]

In the above, there are some echoes of the post-truth notions that were sometimes associated with special relativity.

MANY-WORLDS INTERPRETATION

The concept of the wave-packet collapse troubled many physicists from the earliest inception of quantum mechanics and has spawned many interpretations. One of the well-known alternatives to the Copenhagen Interpretation is the Many-Worlds Interpretation.

It was developed in 1957 by physicist Hugh Everett in his doctoral dissertation. Everett wrote of the Copenhagen Interpretation, "The Copenhagen Interpretation is hopelessly incomplete . . . as well as a philosophic monstrosity."[14]

The Many-Worlds Interpretation attempts to do away with the problem of wave-packet collapse by asserting that there is a "universal wave function" that contains everything. When a wave packet, a part of the universal wave function, interacts with another entity, the interaction creates multiple forks in the universe where all possible states of the wave function persist into the future. Each state exists in its own particular parallel universe that realizes one of many possible outcomes. In this sense the wave packet never collapses, although in one of the branched universes it has the appearance of a collapse.

One challenge to the Many-Worlds Interpretation is how a localized interaction can spawn multiple replicas of distant parts of the universe: stars and galaxies. The parallel universes cannot communicate with each other, making it impossible to compare the ramifications of the paths not taken. Also, according to Everett, it is impossible for the conscious observer to even be aware of the splitting when it occurs.

The Many-Worlds Interpretation often gets explained with a heavy emphasis on human terms. Physicist Sean Carroll commented on the distinction between the Many-Worlds and the Copenhagen interpretations, invoking Schrödinger's cat:

> The situation in quantum mechanics is superficially entirely different. Think of Schrödinger's Cat. Quantum mechanics describes reality in terms of wave functions, which assign numbers (amplitudes) to all the various possibilities of what we can see when we make an observation. The cat is neither alive nor dead; it is in a superposition of alive + dead. At least, until we observe it. In the simplistic Copenhagen interpretation, at the moment of observation the wave function "collapses" onto one actual possibility. We see either an alive cat or a dead cat; the other possibility has simply ceased to exist. In the Many Worlds or Everett interpretation, both possibilities continue to exist, but "we"

(the macroscopic observers) are split into two, one that observes a live cat and one that observes a dead one. There are now two of us, both equally real, never to come back into contact.[15]

Like other interpretations of quantum mechanics, although the Many-Worlds Interpretation is meant to resolve questions at the atomic scale, but the concept has inspired numerous cultural realizations. Like Schrödinger's Cat, the popular renditions portray different outcomes arising from a random event. This was explored in a TV series, *Community*, in an episode titled "Remedial Chaos Theory." In the episode a group of friends gather for a housewarming, with a number of subplots brewing. The characters engage in a game where a six-sided die is thrown to select the person who must go to the front door to collect a pizza delivery. Each outcome of a thrown die results in divergent situations with the subplots, and each outcome is explored as a set of parallel universes. The nature of television is such that each parallel universe must be presented sequentially, rather than in parallel, but one is supposed to infer that they all unfold in parallel.

QUANTUM SUICIDE AND IMMORTALITY

When Everett's thesis was first published in 1957 it received scant attention, but physicist Bryce DeWitt popularized it in the 1970s. In turn, this spawned many new interpretations of the question of wave-packet collapse—an enterprise that has still not settled down.

One curious variant on this theme is the concept of quantum suicide and immortality. This is a thought experiment intended to resolve the differences between the Copenhagen Interpretation and the Many-Worlds Interpretation. As we've seen above, the experiment involves human-scale effects to resolve atomic scale conundrums.

Quantum suicide and immortality is a twist on the Schrödinger cat thought experiment where an observer takes the place of the cat and "observes" his or her mortality by sitting in the box for a period when the likelihood of death is 50 percent. After that period, the

232 * CHAPTER TEN

person is either alive or dead in the Copenhagen Interpretation, but in the Many-Worlds Interpretation there are two universes created, one where the observer is alive and one where he or she is dead. In the second trial, the 50 percent of the living observers tries again, and there are again two branches where there is again a 50 percent chance of survival. Now we have four universes, one where 25 percent of the observer is alive and 75 percent dead, and so on. After many trials, the limit is of a very large number of universes with a dead observer, but one with a living observer.

As with Schrödinger's cat, one cannot take quantum immortality literally, as the coherence lengths and times required for quantum interference does not match human (or cat) scales. Additionally, death itself is not a quantum probabilistic process.[16]

RELATIVITY AND QUANTUM MECHANICS

In this chapter I described some of the elements central to quantum mechanics and some associated social metaphors. All the above is described in a domain that is not relativistic: these are all phenomena that arise in situations where velocities are nowhere near the speed of light. However, Western physics and astronomy have a kind of collective ambition to describe the universe as wholistically as possible at all distance scales. Following this line of reasoning, it would seem natural to marry the two concepts to form a theory of relativistic quantum mechanics.

Both from the side of theory, and from experimental results, the joining of the quantum and the relativistic worlds led to many perhaps surprising results about how we look at space at the very smallest distances.

* 11 *

THE WISDOM *of*
the INWARD PARTS

At the largest scales, the universe has its own rules—the curvature of space and the budget of matter and dark energy. Likewise at the smallest distance scales, the universe operates with its own rules. We approached our explorations of the short-distance scales with certain assumptions about space, but time and again we've been surprised by how nature dictates the "regulation" of physical laws. I am often reminded of this passage in the book of Job, where God appears out of a whirlwind and questions Job on his depth of knowledge.

> Can you bind the chains of the Pleiades, or loose the bands of Orion?
> Can you bring forth Mazzaroth in his season? or guide Arcturus with his sons?
> Do you know the ordinances of heaven? Can you establish their rule on earth?
> Can you lift up your voice to the clouds, that a flood of waters may cover?
> Can you send forth lightnings that they may go and say to you, "Here we are"?
> Who has put wisdom in the inward parts? or who has given understanding to the mind?

(Job 38:31–36)

In this passage the original term for "inward parts" is *tuchah* in Hebrew. It only appears twice in the Old Testament. The other appearance is in Psalms. It's a curious phrase and could be interpreted as meaning the "rules" of the small. In the context of the passage of Job, it's a kind of counterpoint to look at the "rules" of the large with mention of the stars at the start, a terrestrial world described in the middle, including lightning, and then the "inward parts" and the mind.

The first inklings of the rules of the small came in the seventeenth and eighteenth centuries. Famously, there was the curiosity that rubbing fur on amber produces one kind of electric charge and silk on glass another. Ben Franklin was the first to articulate that there were two kinds of electric charge: positive and negative. He speculated that all matter had electric charges associated with it. Franklin is famous for his kite-flying experiment that established that lightning was a phenomenon associated with electricity.

The nineteenth century had many breakthroughs with the development of electricity and the theory of electromagnetism. As we saw, both quantum mechanics and the development of special relativity emerged from considerations related to electromagnetism. There were also dramatic advances in chemistry. Dimitri Mendeleev developed the periodic table of the elements showing the emergence of the rules of atomic interactions, including atomic numbers and atomic mass.

Tests with electric phenomena showed a curious manifestation called cathode rays. These were what seemed to be particles emitted from the negative plate of an electrical circuit in an evacuated tube. The glass in the tube would often glow when current was running. Finally, at the end of the century, physicist J. J. Thomson used a magnetic field to deduce the mass of the cathode rays and found that these were particles about 1/1800 times the mass of the hydrogen atom, which became known as the electron, the first subatomic particle discovered.

Thomson believed the neutrality of atoms came from an intermingling of positively charged particles and electrons in what is sometimes called the plum-pudding model. In this model, the positive charge acts like a uniform pudding with negative electrons interspersed throughout.

Physicist Ernst Rutherford tested this by using alpha particles from radioactive decay as a probe of atoms. The plum-pudding model predicted that most/all of the alpha particles would pass directly through a thin sheet of gold foil, but instead there were some rare very wide-angle scatters. The deduction was that the alpha particles were being scattered by the electric field of a very small nucleus made up of positively charged particles. This would look like a tiny solar system with the electrons being the planets and the heavy nucleus being the Sun.

As quantum mechanics developed it became clearer that electrons weren't like orbiting planets, but rather a wave-like cloud of probability surrounding the nucleus. We call the atomic electron clouds "orbitals."

The further development of particle physics in that era was enabled by the invention of the cloud chamber. Physicist C. T. R. Wilson invented a particle detector that contained purified air saturated with water in solution. When the air pressure was suddenly dropped, water droplets would form around the traces of charged particles passing through the chamber.

When a charged particle moves through a gas, like air, the particles interact with the gas molecules and leave trails of ions—electrons and charged atoms. These electron trails formed tiny cloud-like structures and Wilson was able to record the streaks with a camera. Figure 11.1 shows an image of tracks in a cloud chamber.

I sometimes liken the tracks to the contrails from a jet airplane. Contrails form when the atmosphere is close to saturated with water vapor, and the additional vapor from the exhaust of the jet engines pushes the atmosphere into a supersaturated condition.

Over time, the particle trails were sorted into classifications, as they seemed to come in different distinctive types. The Greek alphabet was used to name each of the distinct contrails/rays. Alpha rays

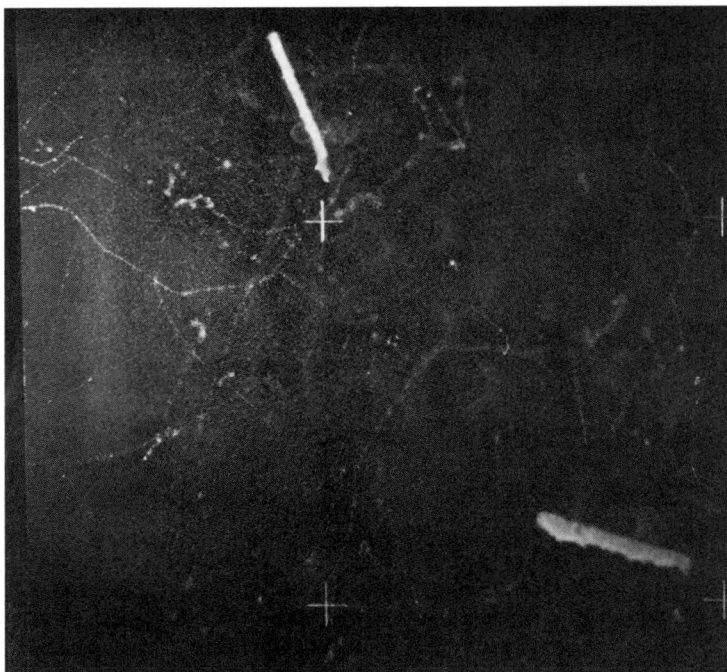

FIGURE 11.1 Photograph of a cloud chamber. The short stubby tracks are alpha particles (two protons and two neutrons). The skinny wandering tracks are beta particles: electrons. The crosses are markers to aid photography. Photo courtesy of Harvard Natural Sciences Lecture Demonstrations.

were thick and relatively short. Beta rays were spindly and wound around, gamma rays were localized bursts. We now know alpha particles are the ion of a helium-4 nucleus (two protons and two neutrons). The beta rays were electrons and the gamma rays high-energy x-rays. All three can come from natural radioactive decay. Figure 11.1 shows both alpha and beta rays.

While atomic structure was slowly becoming understood, other mysteries emerged. First among them was the nucleus itself. If the nucleus was made of protons and neutrons, how did it stay together when we would have naively expected all the protons to fly away from the repulsive electric force? Rather, the nucleus is incredibly compact compared to the size of the atom. The answer is that there must be a force more powerful than the electric force: the strong force. Although the name is not terribly imaginative, the concept is that there is a force that is short-ranged and confined to the distance of the nucleus, only femtometers (10^{-15} meters) in breadth.

Another short-ranged force emerged in the study of nuclear decays that had beta particles, electrons, as an end product. Here, for example, a neutron in a nucleus can decay into a proton, emitting an electron (beta particle). This was and is called the weak force.

In early studies of nuclear decay with electrons as an end product (beta decay), it was found that energy did not seem to be conserved. Without abandoning the core concept of conservation of energy, the idea of a very light, almost massless neutral particle emerged that was believed to take up the unseen energy: the neutrino. The physicist who proposed the neutrino, Wolfgang Pauli, called it a "desperate remedy," and that "I have done a terrible thing, I have postulated a particle that cannot be detected." Not only has it been detected, but we now have observatories for neutrinos coming from astronomical sources.

Together the strong and weak forces explained a major mystery. In the nineteenth century the geological record and evolution pointed to an age of the Earth that was at least hundreds of millions of years old. But there was no model that explained why the heat from the Sun could persist over geologic times. One major hypothesis in the nineteenth century was that its heat came from meteors continually crashing into the solar atmosphere. This idea was championed by physicist William Thompson (Lord Kelvin) who also estimated the age of the Earth to be roughly 20 million years. Another idea in the nineteenth century was that the Sun was slowly contracting and the gravitational energy from matter "falling" inward generated the solar heat. Neither of these could explain the apparent age of the Earth from geology.

With the strong and weak interactions, scientists could now explain what powers the stars. Hydrogen, the most abundant element in the universe, could "burn" in a combination of strong and weak interactions into heavier elements, sometimes terminating in a star-ending supernova. It's worth pausing to ponder that our knowledge of much of the universe comes from observing stars at great distance scales, and yet the source of their power comes from forces that are confined to distances of the scale of the atomic nucleus. Once it was understood where the Sun's power came from, the ages in geologic records could be reconciled with an age of several billion years.

NUCLEAR WEAPONS

Arguably, the most enduring legacy associated with particle physics is the development of nuclear weapons. In 1938 chemists Otto Hahn and Fritz Strassmann, along with physicists Lise Meitner and Otto Frisch, observed that when an isotope of uranium was exposed to slow neutrons, it disintegrated into lighter nuclei and more neutrons. Soon after the discovery of this process of fission, there was a wide-spread recognition among physicists that a sufficiently large quantity of enriched uranium could be used to create a powerful bomb through a chain reaction of disintegrating nuclei and leftover neutrons causing fission until there were no more nuclei left. There was also a possibility of a bomb made with the heavier element plutonium.

Although the knowledge was widespread, the challenge was the vast scale of the industrial production of fissionable isotopes of uranium and plutonium. A secondary challenge was finding ways of compressing the isotopes to enable the critical chain reaction. As is well known, the United States, fearing the Nazis were also on the path of development, embarked on the Manhattan Project to develop deliverable fission bombs. By the summer of 1945, although the Nazis were already defeated, the use of nuclear bombs on Hiroshima and Nagasaki was intended to hasten the end of the war, and to send a message to the world that the United States had developed the weapon. President Harry S. Truman made the decision to use the bombs against Japan. In his speech following the detonation of the first atomic bomb over Hiroshima he declared,

With this bomb we have now added a new and revolutionary increase in destruction to supplement the growing power of our armed forces. In their present form these bombs are now in production and even more powerful forms are in development.

It is an atomic bomb. It is a harnessing of the basic power of the universe. The force from which the sun draws its power has been loosed against those who brought war to the Far East.

The implications of the bomb were immediately grasped, particularly by scientists who worked on the project, many of whom had mixed feelings about their participation. Some embarked on campaigns of nuclear nonproliferation.

One of the most poignant statements on the new weapons was by Albert Einstein in a telegram he sent to several hundred prominent Americans, captured by the *New York Times* in 1946. Einstein appealed for an educational fund, declaring that "a new type of thinking is essential [in the atomic age] if mankind is to survive and move toward higher levels."

But the genie was out of the bottle. Once scientists delivered the knowledge of nuclear weapons to the military and politicians, there was only so much that could be done to rein in the proliferation. By 1949 the Soviet Union tested its first atomic bomb.

Physicists Edward Teller and Stanisław Ulam believed that a far more powerful bomb could be developed by employing fusion of hydrogen, the same process that fuels stars to burn. The specific fusion process they imagined was fusing a deuterium nucleus (one proton and one neutron) with a tritium nucleus (one proton and two neutrons) to form a helium-4 nucleus (two protons and two neutrons) and leaving one free neutron. The bomb would be ignited by a fission weapon that initiates the fusion component and then provides a source of high-energy neutrons that allows an almost unlimited burning of a more inert isotope of uranium. This fission-fusion-fission weapon, that we call the hydrogen or thermonuclear bomb, was a thousand times more powerful than the bombs that were dropped on Hiroshima and Nagasaki, capable of annihilating all the boroughs of New York City. In 1954 the first deliverable hydrogen bomb was tested on Bikini Atoll in the Marshall Islands. It produced a cloud of radioactive fallout that drifted over the atolls of Rongelap, Utirik, and Ailinginae, causing radiation sickness among the inhabitants.

Combined with the development of intercontinental ballistic missiles, the United States and Soviet Union embarked on an arms race, where the consequences of a full-on nuclear exchange between the superpowers could threaten human life on earth. This produced a kind of death pact that was named "mutual assured destruction" or MAD, which inhibited all-out war, but spawned various proxy wars in its stead.

Although the production of nuclear weapons is challenging, it is not out of reach of many industrialized countries. According to the advocacy organization The Federation of Atomic Scientists, there are 9,500 nuclear warheads actively available for deployment by the United States, Russia, China, France, the United Kingdom, Pakistan, India, Israel, and North Korea.[1] As of this writing, Iran is working on uranium isotope separation for reactors, but it also could be used for bombs. The United States's system for launching nuclear weapons is termed "the strategic triad," that could be summarized by where they're based: land, air, and sea. Land-based nuclear missiles are kept in hardened silos. The US Air Force deployed and continues to deploy strategic bombers. The US Navy has stealthy submarines that are effectively submerged missile platforms that can sneak up on an enemy coastline and launch nuclear weapons on short notice.

SPIN AND ANTIMATTER

Be warned, the material that follows can be a bit complicated, as was the case for the beginning of the relativity chapter. Partly, this is because the "rules" of elementary particles are complicated. The "rules" have to be deduced from experiments, like the one I work on at the European Center for Nuclear Physics (CERN) and informed by theoretical models. At some level I wish the descriptions for the fundamental forces were simpler, but it is just the way nature works at the smallest distance scales. The rules of the strong and weak interactions are perhaps the most challenging and I've included these for the interested reader. Perhaps some new theory will emerge that will reduce our model of elementary particles to something more primordial, but at the moment this is what we have.

Classically, when we think of spin, we picture a macroscopic object, like a baseball, or a bicycle wheel that rotates, and the spin maintains a rotational inertia that stabilizes its motion. Perhaps the best example of this is a gyroscope, which is in widespread use as a navigational aid because its spin axis always points in the same direction. In the first half of the twentieth century, physicists discovered that elementary particles possessed a property that is sometimes called "intrinsic spin." To the best of our knowledge the elementary particles we deal with are point-like and aren't extended objects like baseballs, yet they have spin associated with them.

In the same way electric charges on electrons and protons come in discrete units, spin is likewise discrete, coming in units of Planck's constant. Due to their behavior, we have classified elementary particles as fermions, which have a spin ½ or bosons, which have spins of 0, 1, or 2 in units of Planck's constant. Examples of fermions are electrons and protons, while photons, with a spin of 1, is a boson.

In the previous chapter I considered quantum mechanics where special relativity did not come into play. Extensions of that version of quantum theory could accommodate both spin ½ and spin 1 particles readily, but the ambition of science is to provide a wide-ranging framework to describe nature at all scales. A more encompassing theory would incorporate a relativistic version of quantum mechanics.

In 1928 physicist Paul Dirac set out to produce a theory that incorporated special relativity, quantum mechanics, and the properties of spin ½ particles. Here is a case where theory led experiment. Dirac ended up with a workable solution, except there was a problem. His equations predicted a particle with the same mass as the electron, but with the opposite charge. This new particle, called the positron, had not been seen, and Dirac initially thought that his theory was simply incorrect. At first, he associated the positron with the positively charged proton, but the proton has about 2,000 times the mass of the electron. Finally, Dirac relented and had to admit that his theory predicted a particle that had yet to be observed. This was in 1928. Then in 1932, the positron was discovered in cloud chamber photographs taken by the physicist Carl Anderson.

The positron is a kind of particle more broadly known as antimatter. All spin ½ particles (fermions) have an antimatter particle associated with them that has the same mass, but an opposite charge. For example, the positively charged proton has a negatively charged antiproton associated with it.

Antimatter has some unusual properties. When an antiparticle and a particle collide, they annihilate and create a high-energy photon. Figure 11.2 illustrates the annihilation of an electron and positron (anti-electron) into a photon. All the energy of the electron and positron turn into the energy associated with the photon. Physicists Richard Feynman and John Archibald Wheeler discussed the nature of the positron and realized that it could be described as an electron traveling backward in time. In figure 11.2 you can see the arrow

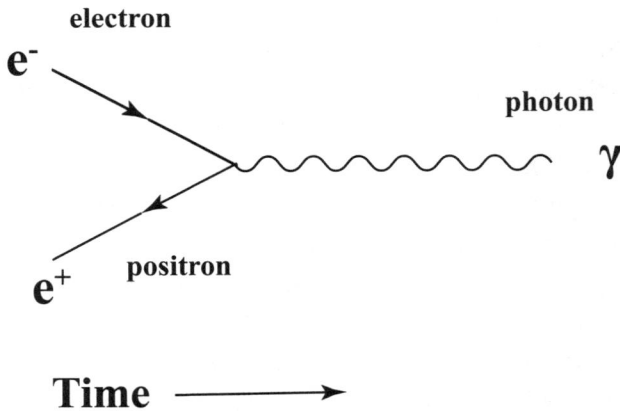

FIGURE 11.2 Electron and positron colliding to create a photon.

indicating our forward direction of time. The electron is shown as traveling forward, while the positron is shown as traveling backward. In a curious way, it would also be interpreted as the electron travels forward, meets the photon from the future and gets turned around to travel backward in time.

When we develop theories of nature, they are often approximations and contain assumptions that are not necessarily valid, but we must start somewhere. These assumptions involve symmetries of various kinds. Take the concept of time. Many, if not most, of the laws of physics suggest that time can move both forward and backward, yet our experience suggests that time only moves in one direction. We often use the physical metaphor of walking to suggest that the future is "forward" and the past is "behind." But, for example, our classical theory of electromagnetism doesn't distinguish the direction of the arrow of time and we must forbid backward time travel "by hand," as this is our experience.

A similar symmetry exists in Dirac's theory of matter and antimatter. It suggests that there should be an equal amount of matter and antimatter in the universe at large, yet it seems that there is only matter and very little, if any, antimatter. We see this observationally on many scales: here on Earth, in the solar system. When galaxies collide, we don't observe the intense photons you would expect from matter–antimatter collisions.

So, what happened? Why is there all matter and no antimatter? Physicist Andrei Sakharov speculated that in the very early universe that was a small asymmetry in the production of matter and antimatter at the scale of one part in a billion. The matter and antimatter collided and formed a huge number of photons, but because there was a bit more matter than antimatter, the residual matter was left over and forms all the visible parts of the universe that we see in stars. What physical process produced this asymmetry? We don't know. We do observe a small asymmetry between matter and antimatter in experiments we've done at accelerators, but most physicists don't believe that these observations are sufficient to explain what we see in the universe at large.

VISUALIZING INTERACTIONS AND VIRTUAL PARTICLES

After Dirac's successful marriage of special relativity, quantum mechanics, and spin ½ particles, the next step was the development of a consistent theory of electromagnetism. One emergent concept in this new theory called quantum electrodynamics (QED) is a virtual particle. This is a particle that participates or mediates an interaction but does not appear in either the initial or final state.

Imagine two canoeists paddling next to each other. One tosses a heavy medicine ball at the other canoe. The first canoe recoils as the medicine ball leaves, and the second canoe itself recoils as the canoeist catches the ball. The canoes before the ball is tossed are the initial state, and the two recoiling canoes after the ball is tossed are the final state, and the medicine ball acts like the virtual particle.

Let's take a simple and hopefully intuitive example illustrated below (fig. 11.3). Say we shoot an electron at another electron. Since they both have the same charge, the force between them is repulsive and they move away from each other.

We could describe this interaction of electron-electron scattering without quantum mechanics or without special relativity, but to have

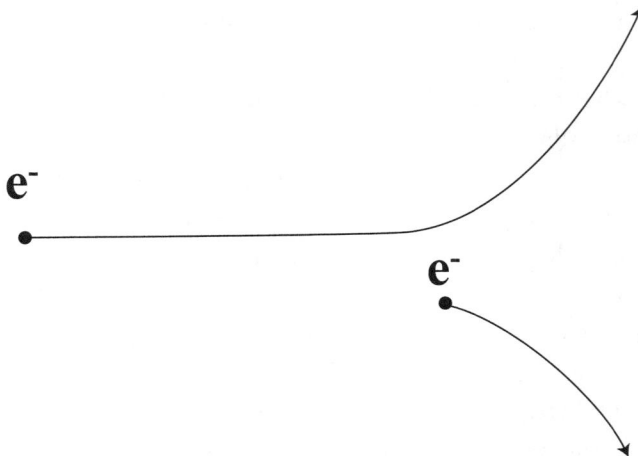

FIGURE 11.3 Electron-electron scattering.

a consistent theory it can be described in terms of quantum electro-dynamics. Classically, we would say that the field lines between the two electrons are responsible for the repulsive force. This action at a distance is also what we see in gravity.

In the late 1940s the theory of quantum electrodynamics was developed. The calculations required had a fair amount of bookkeeping, and physicist Richard Feynman developed a visual device to aid the calculations in what has become known as a Feynman diagram. One element of the Feynman diagrams is the ability to portray inter-actions in terms of virtual particles. While there are real photons, such as x-rays, there can be a virtual photon exchanged between the two electrons that enter the calculation for electron-electron scattering. In this diagram (fig. 11.4), time evolves from left to right and space is represented as the vertical dimension. There are the two electrons in the initial state to the left, before the scattering. The electric force between the electrons is represented by the exchange of the virtual photon, which alters their momentum into the final state, represented by the electrons on the right.

If I take this diagram and rotate space into time (i.e., by 90° from fig. 11.4), we end up with another diagram (fig. 11.5). In this case we

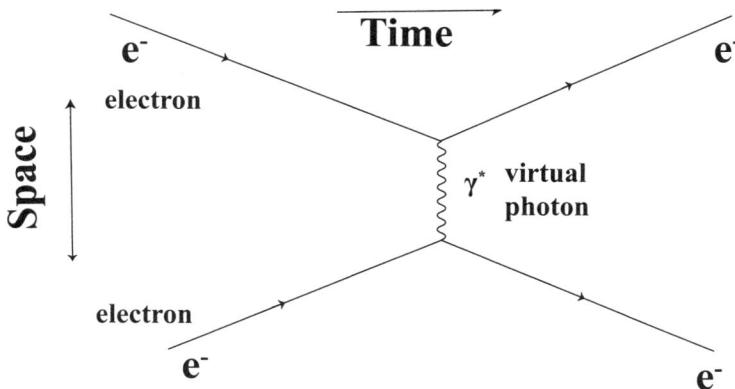

FIGURE 11.4 Electron-electron scattering described in a Feynman diagram with the exchange of a virtual photon between the two particles.

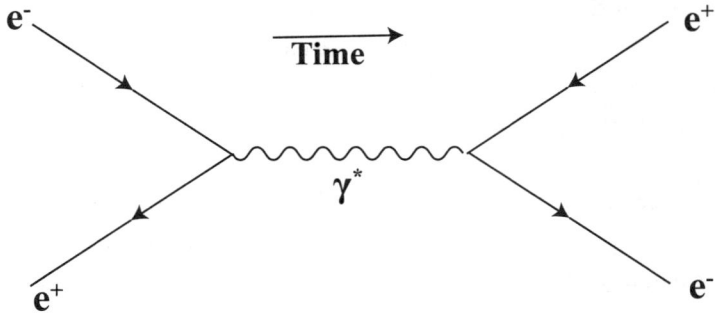

FIGURE 11.5 An electron and positron annihilate and form a virtual photon and then the virtual photon becomes an electron and positron in the final state.

have a diagram describing an electron and positron in the initial state, which annihilates into a virtual photon, which then turns back into an electron-positron pair in the final state.

One advantage of Feynman diagrams is that we can represent more complicated processes. The simple electron-electron scattering diagram shown above (fig. 11.4) gives us an answer to the rate and character of the scatters. But the same interaction can take place if two virtual photons are exchanged. Once we allow multiple particles in the internal virtual states, there are a number of diagrams that enter the calculations. Ultimately, these calculations with more internal virtual states lead to higher precision.

In figure 11.6, there is one important component. If you look at figure 11.7, you can see that there is a case where the photon turns into an electron-positron pair and back into a photon again. This loop is an interesting feature. Normally, we think of free space, a vacuum, as being devoid of any kind of matter. This is a natural way of thinking but, in a sense, space is filled with virtual particles in just the same way that electric fields surround charged particles. In addition to virtual photons surrounding a charge, there are also bubbles of electron-positron pairs surrounding charged particles as part of the cloud of virtual photons.

Here is where we can turn to an important property of space as seen on progressively shorter and shorter distance scales. Say we have

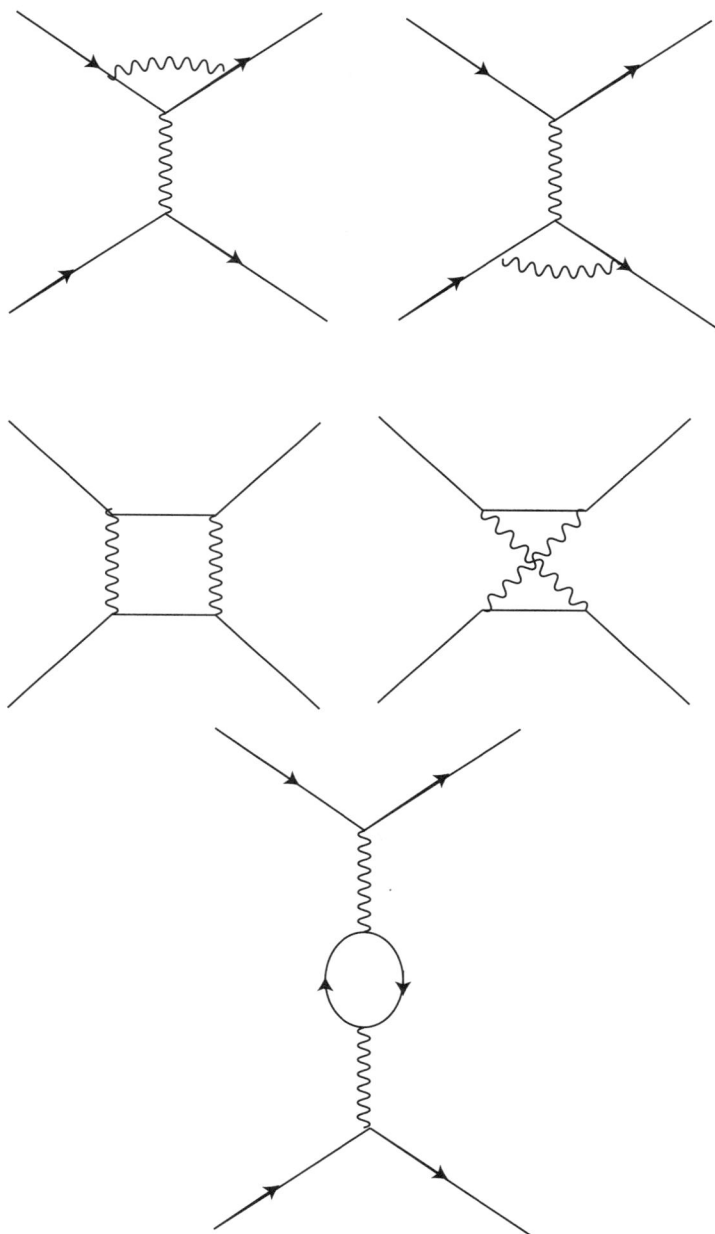

FIGURE 11.6 Electron-electron scattering diagrams with more complicated internal virtual states.

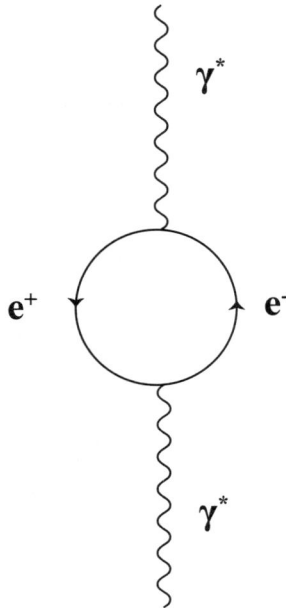

FIGURE 11.7 Focus on the electron-positron loop.

an electron, and we can measure its charge by the way it interacts with other charged particles. We can do this by launching a probe—some kind of charged particle—perhaps another electron. The higher energy the probe particle has, the closer it can get to the electron and see whether its charge is constant or varies with distance.

Here's where the virtual electron-positron bubbles in the vacuum come into play. Those bubbles will feel the charge on the electron. Since unlike-charges attract and like-charges repel, the electron-positron bubbles will orient themselves with the positive charges aligned closer to the electron than the negative charges. This effect can also occur for a charge in matter, like water. The water molecules can get aligned to shield the charge, producing an effect known as charge screening (fig. 11.8). The difference is that this screening isn't happening in matter, it's happening in free space.

From the point of view of the probe that's testing the charge on the electron, the closer it gets to the electron, the more charge it sees

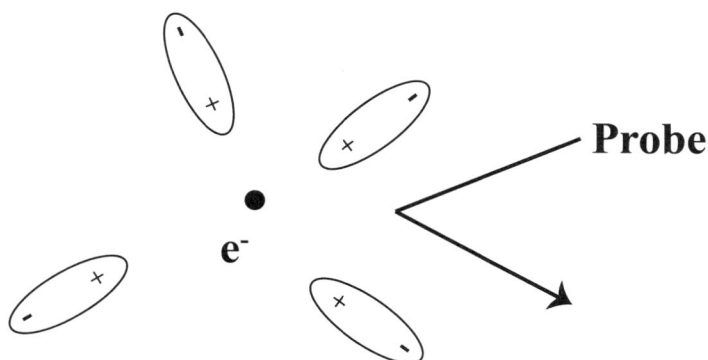

FIGURE 11.8 Charge screening by virtual particles where more and more charge is seen by a probe particle as it gets closer and closer to the electron.

because it's getting more and more inside the screen. As a result, the charge on the electron gets larger and larger as you look at it at shorter and shorter distance scales. Experimentally we see exactly this effect, in accordance with our model of quantum electrodynamics.

It turns out that the strong force behaves in the opposite way. The closer you get to a strong charge, the weaker it gets.

THE STRONG FORCE

The period after World War II into the 1960s is sometimes called the era of the "particle zoo." New detector technologies were deployed. There was the invention of the liquid equivalent of cloud chambers: bubble chambers. There were emulsion stacks—effectively large photographic plates that could record charged particle tracks. High-energy particles from outer space are always colliding with air molecules in the upper reaches of the Earth's atmosphere, creating a shower of particles. By placing emulsions high on mountains, physicists could catch images of new kinds of particles. In turn, new accelerators gave larger energies to create yet more massive particles. At some level this was an embarrassment of riches, but the challenge was to make sense of it all.

Slowly a taxonomy emerged to describe the characteristics of the newly found particles. Broadly speaking, any strongly interacting particle is called a "hadron" from *hadros*, meaning "bulky" in Greek. Of the hadrons, there are two subcategories, baryons and mesons. Baryons, from *baryos*, or heavy, were more massive particles like protons, neutrons, and their heavier cousins. Mesons, from *mesos*, or middle, were heavier than electrons, but lighter than baryons. One of the first mesons to be discovered was the pi-meson or pion, found in 1947.

Over time regularities were seen in the properties of the particles in the zoo: the masses, charges, decay pathways, and a new property called strangeness. By 1961 the classification of these particles led to a schema by physicists Murray Gell-Mann and Yuval Ne'eman that became known as the Eightfold Way. This classification system predicted the properties of a new baryon, the Omega-minus, which was discovered in 1964.

The very existence of this classification system of regularities for mesons and baryons was strongly reminiscent of Mendeleev's periodic table and hinted at a substructure of constituent particles making up the hadrons in the same way the elements displayed regularities. In 1964 physicist George Zweig and Gell-Mann independently proposed that the regularities of the Eightfold Way could be explained by a group of constituent particles now known as quarks. Zweig named them "aces," while Gell-Mann named them "quarks," from a passage in *Finnegans Wake*, "three quarks for Muster Mark." To explain the Eightfold Way regularities, only three quarks were needed at the time, greatly reducing the number of particles considered elementary.

In the ace/quark model, there was an up quark with a charge of +⅔ of the proton charge, a down quark with a charge of −⅓ of the electron charge, and a strange quark with a charge of −⅓.

The existence of quarks presented a conundrum. They seemed to explain a lot and reduced the complexity of the particle zoo. But where were they? We should have seen tons of particles with charges of ⅔ and −⅓. In analogy to the periodic table, when we bombarded nuclei they broke up into lighter constituents, but this was not the case. Whenever the hadrons collided with other particles, only hadrons emerged. But if we used accelerators as powerful microscopes to probe deep within protons, the quarks inside seemed to behave just like free particles, floating around inside the protons. We attribute the lack of free quarks in the wild to a process called confinement.

Our current model of the strong interactions has quarks as spin ½ particles with the fractional charges. The theory is modeled after quantum electrodynamics, and the analog of electrical charge is called color. Rather than one kind of charge, there are three colors, typically named red, green, and blue. The hadrons we see are color neutral. Baryons have red, green, and blue color charges so, at some level, you could say that the baryons are "white" and hence color free. The mesons have color-anti-color combinations like red and anti-red and are also color neutral.

Figure 11.9 shows our classification system for the spin ½ particles, including the quarks. At the time of the Gell-Mann/Zweig model, there were only three known quarks: up (u), down (d), and strange (s). The electron had been around since the time of J. J. Thomson and does not feel the strong force. The neutrino was invented to explain features of beta decay and later found to be a real chargeless and nearly massless particle. The electron has two heavier cousins: the muon and the tau, and their associated neutrinos. Collectively, they're known as leptons.

The force carrier of the strong interactions, the analog of the photon, is called a gluon. Like the photon, the gluon is both massless and is spin 1. However, while the photon carries no electrical charge, the gluon carries color charges. This last point makes a huge difference in the characters of the electrical and strong forces.

If we return to the idea of how charges behave at different distance scales, while electrical charge gets larger as you get closer to it, the color charge gets smaller as you approach it. We can invert the thinking and ask what happens as you move away from the color charges. The electrical charge gets lower as you move away from it, and at an

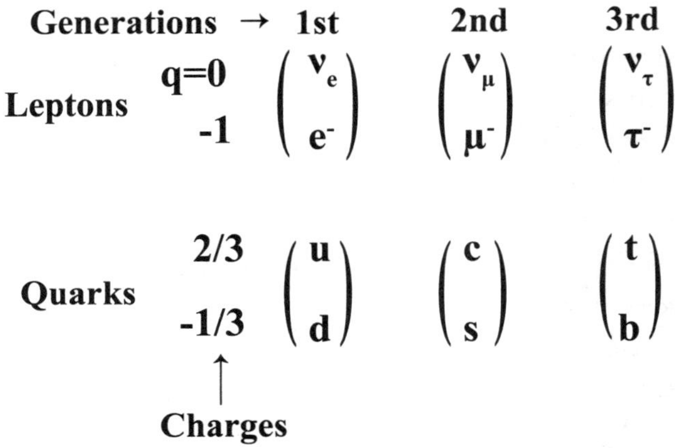

Generations → **1st**   **2nd**   **3rd**

$$\text{Leptons} \quad \begin{matrix} q=0 \\ -1 \end{matrix} \quad \begin{pmatrix} \nu_e \\ e^- \end{pmatrix} \quad \begin{pmatrix} \nu_\mu \\ \mu^- \end{pmatrix} \quad \begin{pmatrix} \nu_\tau \\ \tau^- \end{pmatrix}$$

$$\text{Quarks} \quad \begin{matrix} 2/3 \\ -1/3 \end{matrix} \quad \begin{pmatrix} u \\ d \end{pmatrix} \quad \begin{pmatrix} c \\ s \end{pmatrix} \quad \begin{pmatrix} t \\ b \end{pmatrix}$$

↑
Charges

FIGURE 11.9 The spin ½ particles in the Standard Model. The quarks feel the strong, weak, and electromagnetic forces. The leptons (electrons, neutrinos, etc.) don't feel the strong force. There are three "generations" of the quarks and leptons.

infinite distance it settles down to a nominal value. On the other hand, as you move away from a color charge it gets larger.

So, imagine that we have two quarks, or a quark and an antiquark, close together. Say these are in a color-neutral state, but we give them a big kick to push them away from each other. As they move farther away from each other, they see more, not less charge. The strong force becomes stronger and stronger. At a certain point, the force becomes so strong that quarks and antiquarks get pulled out of the vacuum and form color-neutral states, like mesons and baryons.

This process explains why we see no free quarks. The process of neutralizing the color charge as the quarks move away from each other is called *hadronization*. Hadronization is not well described by the Feynman diagrams, and we have somewhat ad hoc models for that phase of the strong interaction. Bearing that in mind, at short distances the rules we use for Feynman diagrams seem to work well.

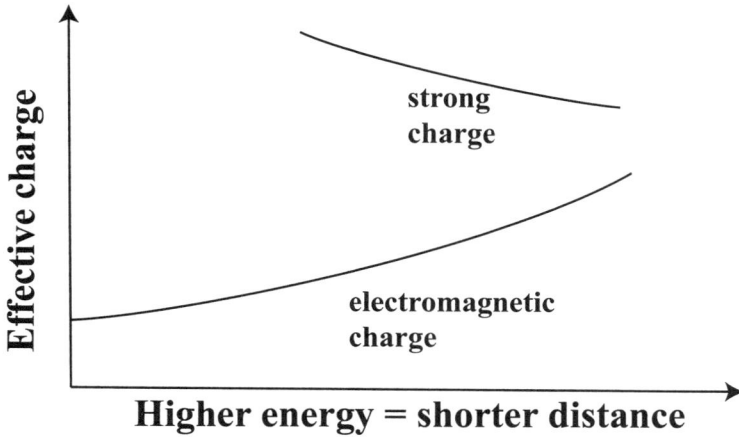

FIGURE 11.10 Electrical charges get stronger at shorter distances, while strong charges get weaker.

Figure 11.10 illustrates how the electrical and color charges scale with distance and energy. Again, the electrical charge gets larger at shorter distances, while the color charges get weaker. You can also note that the strength of the strong charge is greater than the strength of the electrical charge. This, in part, is why the nucleus is held together, rather than flying apart from electrical repulsion.

This feature of the charges changing with distance has been, in part, an inspiration for a more ambitious program in particle physics: a possible unification of the strong and electrical forces at a sufficiently short-distance scale when the charge strengths approach each other and meet.

THE WEAK FORCE

The decays of nuclei into electrons (beta particles) were a sign that there was another short-ranged force that was neither electromagnetic nor strong. Figures 11.11, 11.12, and 11.13 use the decay of a neutron

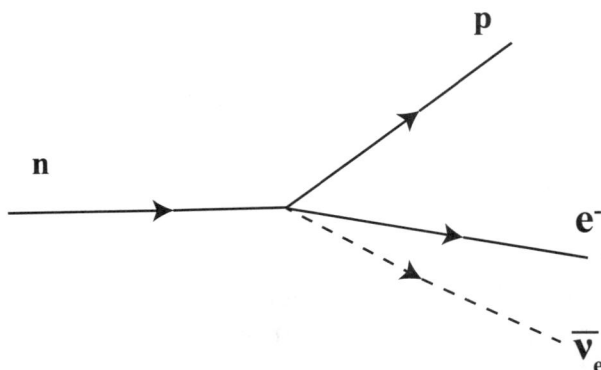

FIGURE 11.11 Fermi's original model of beta decay shown for a neutron decaying into a proton, an electron, and an antineutrino.

into a proton, electron, and neutrino to illustrate the evolution of our understanding of the weak force. This sequence of diagrams is like using a microscope to zoom more and more in on the process of beta decay at shorter and shorter distance scales. In figure 11.11 it looks like the decay all happens at one unresolved point in space. The bar over the neutrino symbol, $\overline{\nu}e$ indicates that it is an antineutrino. This was the first model of beta decay put forth by physicist Enrico Fermi. The neutron turns into a proton and emits both an electron and an anti-neutrino. All four particles have spin ½.

The lifetime of the neutron is about fifteen minutes and is consistent with Fermi's model. It worked reasonably well at low energies for a variety of processes, like nuclear decays, including the neutron. However, Fermi's model had some mathematical pathologies that are beyond the scope of this book. Suffice it to say that it was really considered only a provisional low-energy model, with a more correct model to follow.

Now, here is an important point about virtual particles. In the electromagnetic interaction and the strong interactions, both the photon and the gluon are massless. One feature of the weak interactions is that its force carriers have mass. As it turns out, the range of massive virtual particles is limited in spatial extent. The empirical fact that the weak interactions only exist on the distance scale of the nucleus suggest that the weak interactions are indeed mediated by a massive particle. Another empirical fact about the weak interactions is that many of them exhibit a change in charge. In the case of the neutron decay, you can see that the neutron (charge 0) is changed into a proton (charge +1). These two aspects in the weak decay point to a massive-charged particle that participates in the decay. By convention, it is labeled the W⁻. Like the photon and gluon, it also has a spin of 1.

For our example of the neutron decay above, the next diagram (fig. 11.12) shows a charged W⁻ particle as the mediator or virtual particle that participates in the decay. The neutron connects to the massive W⁻ and turns into a proton. The W⁻ subsequently couples into the electron and anti-neutrino, giving the same final state as the Fermi model. As a sanity check, we can look and compare the charges in the initial and final state. Both have zero charge. The neutron (charge 0) turns into a proton (charge +1) and a W⁻ (charge −1) and ultimately into an electron (charge −1) and a neutrino (charge 0).

With the discovery that hadrons are made of quarks bound together by gluons, the theory of weak interactions had to take into account that it is the quarks themselves that are coupling to the W. Figure 11.13 is a kind of magnification again of the neutron decay, this

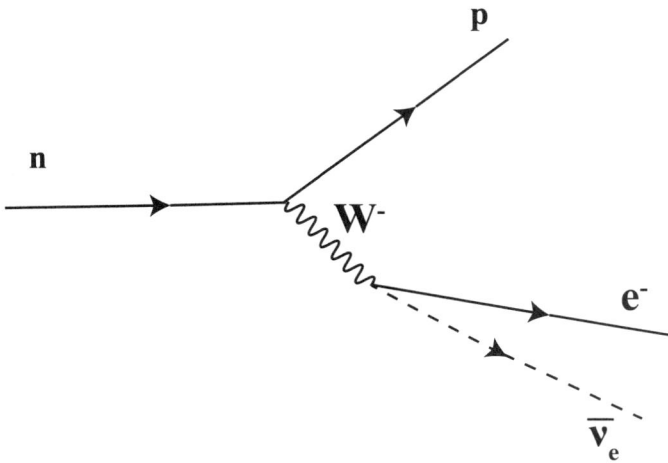

FIGURE 11.12 Neutron decay with a virtual massive W⁻ particle.

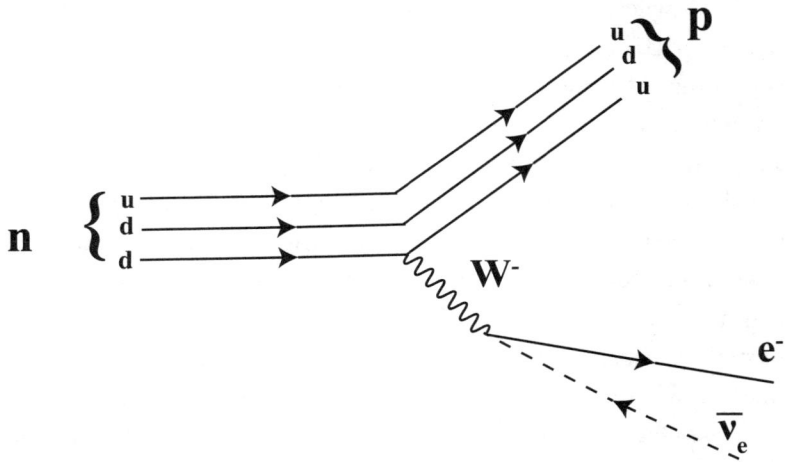

FIGURE 11.13 Neutron decay showing the quark
content of the neutron and the proton.

time showing the quark content. The neutron is made of two down
quarks and one up quark. In the decay process one of the down quarks
turns into an up quark when it connects with the W^-, thus changing
the neutron into a proton with two up quarks and one down quark.
The up and down quarks that don't participate in the interaction are
sometimes called "spectator quarks."

Up until now, I've avoided talking about non-strong charged particles beyond the electron to keep it simple. But nature is more complicated. The electron has a heavy cousin called the muon, denoted
by the Greek letter μ (mu). It was discovered in 1936 and has similar
properties to the electron. The muon is also spin ½ and also does not
feel the strong force, but it has 207 times the mass of the electron. It
came as a surprise at the time, leading physicist Isador Rabi to quip,
"Who ordered that?" The muon came along with its own species of
neutrino, the muon neutrino or ν_μ.

In the above example, the W^- is changing the variety of quark from
a down to an up quark and also connects between an electron and an
antineutrino. Likewise, a positively charged W^+ can participate in the

weak interactions. The W^+ can, for example, change an up quark into a down quark. As far as we know, there are no processes that change leptons into quarks or vice versa, although people have looked.

This coupling of up-to-down quarks and electron-to-neutrino suggests a kind of pairing of the leptons and quarks. The quark model of Gell-Mann and Zweig only needed three quarks to explain the Eightfold Way: up, down, and strange. The pairing of the quarks and leptons, however, suggested that the strange quark must have a partner, the charm quark. This was discovered in 1974 and went along with the lepton partners of the muon and muon neutrino.

But wait! There's more! Subsequent experiments revealed the existence of a heavier pair of quarks, the top and bottom quark. These come along with another two leptons: the tau lepton, and tau neutrino.

Returning to figure 11.9, this shows the full structure of the quarks and leptons as contained in what we call the *Standard Model*. The leptons and their charges (q) are listed on the top rows, and the quarks and associated charges are listed on the lower rows. Each pairing of quarks and leptons are given a designation of "generation" (e.g., the up and down quarks belong in the first generation).

INTERLUDE: LIFE ON EXPERIMENTS AND THE TOP QUARK DISCOVERY

I hope this section will give the reader some insight into the workings of modern particle physics. First and foremost, we are in the age where many discoveries in particle physics come from large accelerator-based experiments.

I currently work on an experiment called ATLAS at the European Center for Nuclear Physics (CERN) at an accelerator called the Large Hadron Collider (LHC) that collides protons at very high energies. ATLAS approaches the size of a Medieval gothic cathedral at 46 meters long and 25 meters in diameter. It is designed to record the remnants of ultra-high-energy proton-proton collisions at the center of the detector. Over 100 million channels of electronics record the results of a billion collisions a second. Across the LHC from ATLAS is our sister experiment, CMS, which has roughly the same scale as ATLAS and was designed to explore similar physics, with some differences in its construction. Together, the two experiments provide the kind of redundancy we want in science.

Building and operating such a complex experiment can be a real challenge. I joined the ATLAS collaboration in 1995. I wouldn't call it an "assignment," but I signed up to help with the part of the experiment that detects muons. In particular, I worked on the design and production of approximately 500,000 channels of electronics. The initial design and testing work required many trips from Harvard to Geneva. In the meetings at CERN we hammered out the detector and electronics requirements. There are multiple restaurants at CERN and frequently details were discussed over an espresso, a beer, or glass of wine. If you casually listen to conversations in the cafeteria, although English is the common language, you'll hear German, Italian, Greek, Chinese, Hebrew, and more—all the languages of the international collaborators.

There's on-site housing at the CERN Hostel. There is a minor problem: if you stay in the hostel on a long visit, you may never leave the CERN campus, and this can get humdrum real fast. Trams now can take you into downtown Geneva. To avoid the monotony I often stay off-site in a hotel and have dinner with my colleagues and students at local restaurants. If I visit a favorite pizza place in the nearby village of Meyrin, I will often run into other refugees from the CERN cafeteria. The travel back and forth between the United States and Geneva can be taxing, and as I said in the previous chapter one of my colleagues declared that his wave function peaked over the Atlantic Ocean.

When not physically meeting at CERN, we make heavy use of video conferencing. We are so accustomed to video meetings that when the COVID-19 pandemic hit, the switch to all things virtual was not so difficult. But it did reveal one facet of being present at

CERN. When you're there, you're constantly bumping into people you know, and you can go grab a coffee and discuss some matter related to the construction of the detector, its operation, or a data analysis topic. After the COVID-19 quarantines lifted, we slowly got back into the routine of periodic trips to CERN. Many graduate students and postdoctoral scientists live in the area, and they are the true front-line soldiers who know how to operate the detector.

The discovery of the top quark provides a window into the working of particle physics experiments and collaborations. The top quark is the most massive of any of the particles in figure 11.9. To produce more massive particles, more energy and/or beam intensity is required for their production. Roughly speaking, the energy required for particle production can be understood from Einstein's famous $E=mc^2$. More energy is required to produce a higher mass particle.

Physicists typically report masses in terms of energies. We use a unit called a GeV for Giga-electron-volt for energies. For reference, the proton has a mass of close to 1 GeV. By the time of Gell-Mann and Zweig, the up, down, and strange quarks, which are relatively light, had been realized from the Eightfold Way. In 1974 the charm quark was discovered simultaneously by two experiments, one on the East Coast and one on the West Coast of the United States. The charm quark itself has a mass of roughly 1.3 GeV, a bit over the mass of the proton. This is a little misleading, as the production of the charm quark comes in pairs, so something like twice the mass energy of the charm quark was required to produce it.

In 1977 and 1978 the b (or bottom) quark and the tau lepton, the third lepton after the electron and muon, were discovered. This implied that there was a third generation of quark and lepton pairs. To fill out the table of quarks and leptons, the tau neutrino and the top quark needed to be found. The mass of the b quark is roughly 4 GeV.

It's difficult to assign a mass to the up and down quarks, but they're very light. The mass of the strange quark is sometimes quoted as 0.1 GeV, but that's also a little challenging to defend. When quarks get heavier, their mass is a bit easier to pin down. So, with 0.1 GeV for strange, 1.3 GeV for charm, and 4.2 GeV for bottom, it seemed like the discovery of the top quark was just around the corner.

Two accelerators were constructed and started running in the early 1980s: PEP at the Stanford Linear Accelerator Center, and PETRA at the German accelerator laboratory DESY. Both were accelerators where electrons and positrons were collided and had beam energies of roughly 15 GeV. Many felt at the time that the top quark was within reach, but neither accelerator produced any hint of the top. I did my thesis work on data taken by an experiment called the TPC at PEP. It was a search for fractionally charged particles, presumably free quarks. I didn't find any.

In the mid-1980s there was an accelerator at CERN, the SppS, which was involved in the direct production of the W boson, and its neutral partner, the Z. The SppS collided protons and antiprotons. In 1984 one of the experiments, UA1, reported the discovery of the top quark at 40 GeV. Their sister experiment, UA2, did not confirm this and it later turned out that UA1 had not properly estimated the backgrounds that could mimic a top quark at that mass. So, ultimately the report was wrong.

After the SppS, the next large accelerator to come online was the Tevatron at Fermilab in Batavia, Illinois. Like the SppS, the Tevatron collided protons and antiprotons together, but at higher energies. Each beam at the Tevatron was roughly 1000 GeV. Now, this doesn't translate directly into a mass reach of 1000 GeV. The proton and antiproton are made up of quarks and gluons, and these carry some fraction of the energy of the proton and antiproton. Really, it's the energies of the quarks and gluons that can directly couple to the production of a heavy object.

Figure 11.14 shows two possible production mechanisms for a top and antitop quark in proton-anti-proton collisions. In the first, a quark from a proton and an antiquark from an antiproton annihilate to a gluon (corkscrew-like figure), which then turns into a top and antitop pair. In the second, gluons from the proton and antiproton connect to a top and antitop pair.

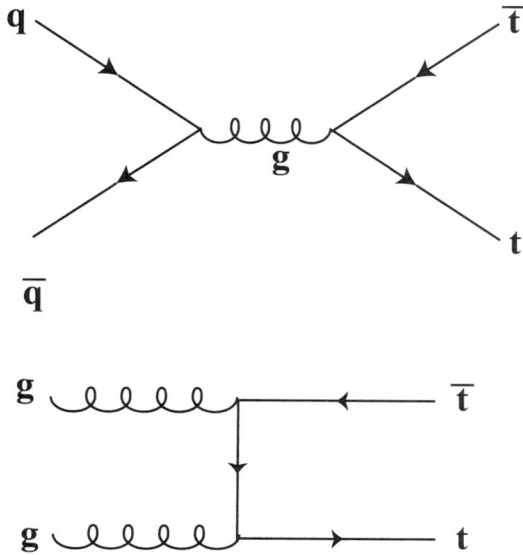

FIGURE 11.14 Two contributions to top quark
production in proton-antiproton collisions.

Once a top quark is produced, it decays in much the same way the
neutron decays into a proton via beta decay. The top quark turns into
a bottom quark and a W. The W proceeds to decay into any number
of possible final states, in figure 11.15, for example, a positron and an
electron neutrino or a quark and an antiquark. I should note that this
figure is relevant for a very heavy top quark, where the W in this case is
a real particle, not a virtual particle. Early on at the Tevatron, we ruled
out lighter variations of the top where the W wasn't a real particle,
so this figure is relevant for our search that uncovered the top quark.

Since there is a top and antitop quark produced in the above diagrams, we'll have two b quarks in the detectors, and then two Ws. Each
W then decays into any one of a number of possible final state particles.
This could be an electron plus neutrino, or muon plus neutrino, a tau
lepton plus neutrino or a pair of quarks. So, a signature for the production of a top quark and an antitop quark might be two b quarks, an
electron, a muon, and two neutrinos. Another possibility might be two

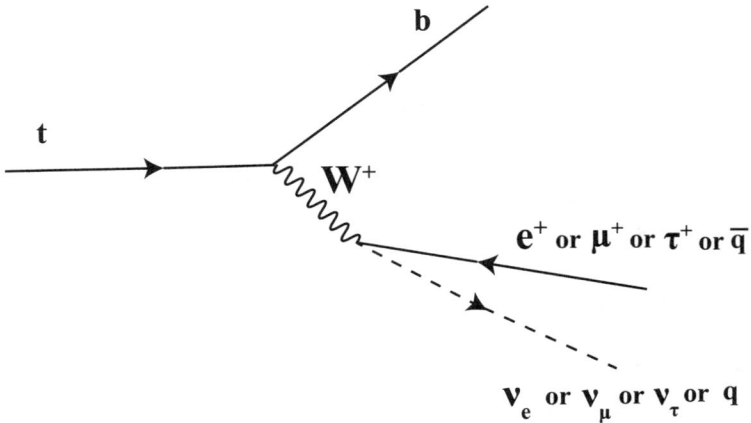

FIGURE 11.15 Decay of the top quark into a b quark and a W. The W subsequently decays into any pair of final state particles indicated.

b quarks, two other quarks, a muon, and a neutrino. All combinations are possible, although there should always be two b quarks.

So, how do we realize this in an experiment? Figure 11.16 is a sketch of a typical modern experiment found in one of the large accelerator facilities. I'll try to describe how we identify each of the particles in turn. The detectors are a bit like layers of an onion with one layer surrounding another. The beam itself is directed through a tube held under a vacuum in the middle of the detector. At the Tevatron the beams consist of bunches of protons and antiprotons that periodically collide at the center of the detector.

Immediately surrounding the beam pipe is a tracking chamber. In the same manner of a cloud chamber, all charged particles leave an ionization trail in a gas mixture. The main difference here is that the ionization trails are detected by electronic detectors. The tracking chamber is immersed in a magnetic field that bends the charged particles' paths. The bending allows us to find the momentum. The straighter the track, the higher the momentum.

Here is an important point. Recall that I wrote above that we've never seen a free quark, but I just said that one signature of a top quark decay is a pair of quarks. This would seem to be a contradiction, but

I haven't told the whole story. The quarks, as they emerge from the collision, reassemble themselves into a spray of color-neutral objects: mesons and baryons, called a *jet*. If we measure this somewhat collimated spray of hadrons, we can reconstruct, to a good approximation, the energy and direction of the quark that produces the jet.

Next, we can consider the b-quark signature. This will also appear as a jet, but there's an interesting feature. The b quark can be bound inside a meson, which has a finite lifetime. On average it might travel some distance from the initial proton-antiproton collision. If we trace back the particles associated with a b-quark jet, they appear to originate at a point displaced from the collision point. This is labeled as a secondary vertex in the figure.

Next, we can look at how we identify electrons. Electrons are light and if they come near a heavy nucleus, it will feel the electric force, get accelerated, and produce a high-energy photon. If we have a stack of heavy metal, typically lead, interleaved with particle detectors, we

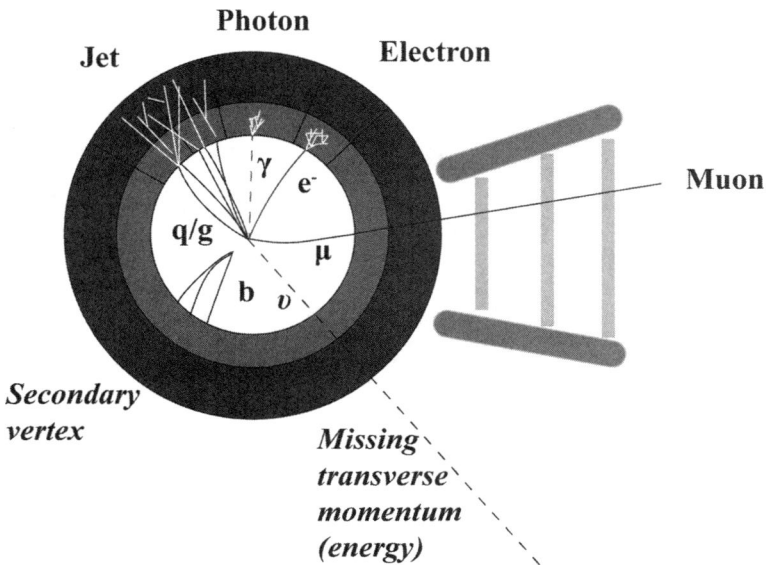

FIGURE 11.16 Schematic of a modern particle physics detector and the signatures of the various particles we find in the final state.

can produce a phenomenon known as an electromagnetic shower. The initial electron will be deflected by a lead nucleus and emit a photon. The photon then can turn into an electron-positron pair, which in turn can release more photons. The process continues until all the energy in the original electron gets dumped into the lead plates. We can measure the amount of energy by looking at the energy sampled in the detectors interleaved between the lead plates. This device is called a calorimeter. Because it is looking for electrons or photons, it's called an electromagnetic calorimeter.

A high-energy photon will also produce a shower like an electron. The way we distinguish a photon from an electron is by looking for a charged particle track in front of the shower. If there's a high-momentum charged track, it's an electron. If there's no track, it's a photon.

Surrounding the electromagnetic calorimeter is a larger calorimeter made out of steel and other particle detectors. The particles that feel the strong interactions, the hadrons from jets, will interact with the iron nuclei and produce their own showers of strongly interacting particles that will dump all their energy into this larger calorimeter, called a hadronic calorimeter.

That brings us to the muon, the heavy cousin of the electron. Like the electron, it does not feel the strong force. Unlike the electron, it's much heavier. This last point is important to identifying a muon. Since it's heavy, it doesn't easily get deflected by a nucleus like the electrons do. We have very energetic particles from outer space, cosmic rays, hitting molecules in the upper atmosphere all the time. The atmosphere acts like a huge calorimeter, and the electrons and strongly interacting hadrons dump all their energy in the upper atmosphere. Almost none of that energy reaches sea level. But the muons do because they don't produce showers. In a typical room, the size of a bedroom, there may be hundreds of muons passing through every minute. The strategy for finding muons is to just look for any charged particle that makes it out of the calorimeters in the detector, so there are specialized muon detectors surrounding the entire detector and anything that makes it out is tagged as a muon.

Last, but not least, are the poor neutrinos. Neutrinos don't have charge, so they don't feel the electromagnetic force. Neutrinos also don't feel the strong interactions, so they won't interact in the calorimeter. In fact the neutrinos will hardly interact at all. Neutrinos will carry energy and momentum, however. The strategy in finding neutrinos is to invoke conservation of energy and momentum. We add up all the energy and momentum in the detector. It all must balance because of conservation of energy and momentum. If there is a lack of balance in energy/momentum, we call this "missing energy" and take it as a sign of a neutrino.

Neutrinos can be made to interact, but it takes a huge amount of material to have a decent chance of seeing them. There's an experiment at the South Pole called IceCube that uses a cubic kilometer of ice to create enough atomic targets to detect the neutrinos from their interactions. Light emitted from the charged particles generated by the neutrino interactions can be collected on a giant array of phototubes. Plans are underway to upgrade the experiment to eight cubic kilometers.

That brings me to the detector proper. I worked on a detector called CDF at the Tevatron. CDF means "Collider Detector at Fermilab." When constructed, it was the size of a large house and had millions of channels of electronics. You can imagine the amount of effort that goes into designing and building one of these things. There's the tracking detector, the electromagnetic calorimeter, the hadronic calorimeter, are the muon detectors, the magnet, all the electronics, and the computing software for reconstructing events. That's a lot.

The detectors are designed, constructed, and operated by large collaborations. As of this writing, there are approximately 6,000 people on my current experiment, ATLAS at CERN, including physicists, engineers, students, and support staff. They come from 257 universities and national laboratories from 42 countries. At the time of the discovery of the top quark, the size of the CDF was smaller, but still large in scope.

Typically, each institute (university or national laboratory) secures its own funding from a national funding agency. The institutes then negotiate with the collaboration to find a task, like a contribution to the construction, installation, and operation of one or more of the detectors on the experiment. The collaboration itself has an internal management structure and bylaws that govern how they operate. There's some combination of a top-down bureaucracy, but also a degree of independence. Navigating all of this can be challenging and one encounters all kinds of individuals with varying skill sets and, indeed, varying people skills.

A person wishing to operate on one of these experiments has to come from a member institute and has to gain the right to analyze data by first doing some kind of service task as an entry. The author lists for scientific papers contain the name of every person on the experiment. The lists have become so long that they are now omitted from publications, and the name of the experiment carries a footnote that points to the author list.

Once a person has gained permission to perform an analysis on data, they usually aggregate into analysis groups that address specific topics. There's a fair amount of flexibility and independence in this choice. Many times it depends on the tastes of a graduate student and their thesis advisor, or a postdoctoral research scientist. There is overlap where multiple groups tackle the same analysis task. Once an analysis group feels that an analysis is ready for publication, they signal their intention to the collaboration, and an oversight group is created to go over their analysis, supporting documentation, and proposed publication. Once the analysis has crossed all the hurdles, it is ultimately submitted for publication in a refereed journal.

There are often groups that specialize in the identification of the various observables. There may be a jet group, a muon group, an electron group, and so forth. These groups typically will standardize the means to identify each of the observables and the people doing analysis will draw upon these standards.

Most analyses will have to cope with backgrounds from sources that can imitate the physics of interest. These must be modeled and accounted for. The lack of proper background modeling can result in spurious claims of new physics, like the UA1 top quark announcement. Sometimes there are simply large statistical fluctuations that can imitate new physics. Invariably the signatures of all kinds of physics must be simulated for the processes themselves and how the detector responds. Comparison of the data to the simulations is the acid test of an analysis. Additionally, real data adjacent to the relevant data can be used as a cross-check on the validity of the simulations. As you can imagine, all this takes a considerable amount of time and computing power.

Several independent groups in the CDF top search were looking at various signatures. Some were looking at the channels where there was an electron, a muon, missing energy, and two b-jets. This was a relatively clean channel, but the existence of two neutrinos complicated matters. One of the earliest hints of a top quark came from these electron-muon events.

Many in the collaboration viewed the most promising channel as the mode where one top decayed into a b-jet, electron, or muon on one side, and b-jet plus other jets on the other side. The decay of the b-jet and electron or muon would include a neutrino, so there would be missing energy. In this "lepton plus jets" channel, there would be a lepton, one or two b-tagged jets with a secondary vertex, two other jets and missing energy.

In this latter class of events, there were several groups with different algorithms for identifying b-jets vying for primacy. Another strategy was to look at the configurations of jets in the events. One large background was the production of a W that recoiled against garden-variety jets. The configurations of these events had a different structure from what was expected from top and two groups found evidence of this difference, although these groups were viewed a bit as outsiders by other members of the collaboration. Nevertheless, it was an important piece of the puzzle.

Ultimately, I was assigned to be one of three arbiters of the analysis. We were called "godparents." The challenge was to convince the groups and collaboration to arrive at a common model that employed all the techniques at our disposal.

Most large accelerators have at least two all-purpose experiments. Reproducibility is a key in scientific endeavors. In the case of the Tevatron, our competition was called the D0 (D-zero) experiment, that had somewhat different approaches to their detector technologies, but had the same key components I outlined above. Relatively early on, D0 also observed one electron-muon event that was suggestive of a top.

By 1994 CDF had a close-to-compelling case for a top quark, and we published an "evidence for" paper for a top quark with a mass of 175 GeV. It wasn't iron-clad in terms of the uncertainties but was getting there. D0 did not report a similar "evidence for" at the time.

More data came in, and we worked with the various analysis groups to form a cohesive and compelling story. This, combined with higher statistics, boosted us to the point of declaring a discovery of the top quark. By this time D0 also had the data and analysis in hand to claim a discovery. In 1995 the two collaborations jointly announced the discovery of the top with a mass of 176 GeV.

As a coda, I should remark that my current experiment, ATLAS, is using our Large Hadron Collider at CERN as a kind of top-factory, where all manner of processes involving top quarks are under study.

Ultimately, it is odd that the top quark is so heavy compared to the other quarks. The spectrum of the masses of the fermions—leptons and quarks—is something of a mystery, although the pattern of having heavier and heavier particles in the higher generations has some consistency to it.

TOWARD UNIFICATION: THE HIGGS BOSON

An ambition of physicists has been a unified description of nature at all distance scales with a single theory. One example of this ambition for a unified theory is a model by Theodor Kaluza and Oskar

Klein, who developed the concept of a unified theory of gravity and electromagnetism in the 1920s by adding an extra dimension of space that was curled up into a tiny circle. Work on the theory continues to this day, but experiments have not turned up any evidence for the small extra dimensions.

The combination of relativity, quantum mechanics, and electrodynamics is a major success and represents a unification of sorts. When I say "success," I mean that quantum electrodynamics combines these three theories into one and made very accurate and precise predictions that have been confirmed experimentally. In the late 1950s it was realized that the weak and electromagnetic interactions were a manifestation of a single interaction dubbed the electroweak force. The unification needed one additional force carrier, the Z°, which is a heavy neutral version of the photon. The four force carriers, the photon, W^+, W^-, and Z°, along with the quarks and leptons could account for electromagnetic and weak interactions in a single unified theory. At present, the strong force stands separate, although there are theories to include this as well.

Focusing in on the electroweak interactions, once it was realized that massive versions of the photon could account for the observed weak interactions, some inconsistencies in the theory were realized. I'll try to sketch these here.

As mentioned in the previous chapter, quantum mechanics gives results that are manifest as probabilities for observable quantities. One important corollary is that the probabilities must be bounded by a kind of sensibility. If we throw a die, we expect a one-sixth chance of it landing on one of the sides. The sum of the probabilities adds up to 1. It doesn't make sense to talk about a probability greater than 1, even with a loaded die. The same is true for quantum mechanics—the sum of the probable outcomes of measurements must be 1. This principle provides an important litmus test for theories called unitarity.

One problem with the electroweak model as of the late 1950s is that some interactions can violate unitarity. Figure 11.17 shows one example of an interaction with this problem. It's the scattering of two

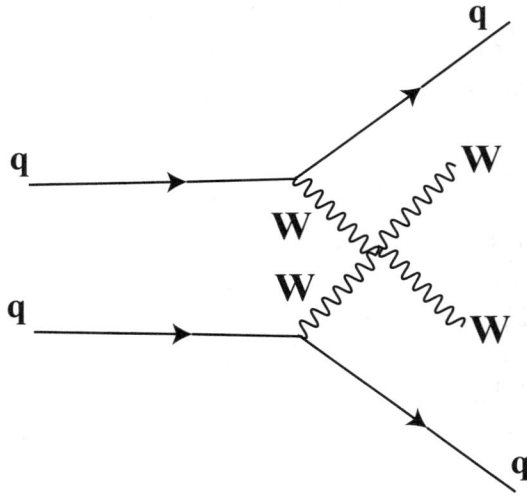

FIGURE 11.17 W-W scattering as seen in a proton-proton collider with quarks from each proton producing a W.

W bosons. This would come from proton-proton collisions where two quarks each contribute a W boson that interact with each other. This is not just an imaginary process for a theorist to contemplate; it's a real process that is observed at the Large Hadron Collider at CERN. In the incomplete theory of electroweak this interaction breaks unitarity at high energies. The problem has to do with the mass of the Ws. There is a component of the spin of the Ws associated with its mass that grows with energy. This becomes so large that the probability exceeds unity in the scattering process.

There was another problem with the electroweak theory associated with the massive Ws and Z. Recall the change in the electric charge with distance scale from the electron-positron bubbles. In the model of quantum electrodynamics, this is well behaved. If you try to do the same thing with massive Ws and Z this process is not well behaved and gives impossible results.

What's the solution to the problems? The mass of the Ws and Z are the problem. If they were massless like the photon, the theory would work perfectly well. The solution is that there is a primordial version of the theory where the Ws and Z are massless, and there is an interaction that gives mass to them.

This brings us to the Higgs boson, sometimes referred to as the "god particle" by journalists in a perhaps unfortunate phraseology. The above problems were understood in the late 1950s/early 1960s. The proposed solution came from an unexpected phenomenon: superconductivity.

In metals electrical current flows with electrons traveling under the influence of an electric field. The electrons will bump into atoms and scatter in different directions. This heats up the metal, and the current will flow with resistance. In the early 1900s it was found that when metals are cooled to near absolute-zero temperatures, electrical resistance vanishes. This state for metals is called superconductivity. Normally, the electrons bump around on their own rather randomly, but when the metal is cooled below a certain critical temperature they become linked into a single coherent quantum mechanical wave.

Now, if I were a reader, I would probably be wondering, "What on earth does this have to do with particle physics?" Here's the catch. If you take a normal, everyday photon and put it in this superconductor, it gains a mass. Does this mean that our theory of electromagnetism is wrong? Not at all! The interaction of the photon with the superconductor is what gives the photon the mass.

Physicist Phil Anderson realized that the problems with the massive Ws and Z could be solved if the universe was somehow a vast superconductor. In that case, the Ws and Zs would really be massless, and the interactions with this superconducting field of the universe would generate their masses, and all would be good.

Cuing off Anderson's suggestion, three physicists developed a full model using this superconductor idea. The most famous of the three was Peter Higgs, whose name we associate with the phenomenon, but physicists Francois Englert and Robert Brout are also credited with the proposed mechanism. One major prediction of this part of the theory is the existence of a particle that has become known as the Higgs boson.

With the Higgs boson and its interaction with the Ws and Z, the theory was well behaved in every way and showed a true unification of the weak and electromagnetic interactions. There was no prediction for its mass, however. There was only a range of possible masses for the Higgs. In addition, it would require a huge accelerator to produce it. In the early 1980s there was talk of a "World Machine"—an accelerator so large that it would take a worldwide collaboration to build it. CERN had a tunnel with a circumference of twenty-eight kilometers that was being used for electron-positron collisions. It was believed that this tunnel could be the home of a large proton-proton collider dubbed the Large Hadron Collider (LHC).

During the Reagan administration, the United States decided to go it alone and embarked on the construction of the Superconducting Supercollider in Texas to go after the Higgs. However, cost increases coupled with cries to rein in the budget deficit in Congress forced its cancellation in 1993. Ultimately, the United States joined CERN in the construction of the LHC and associated detectors. At some level, the LHC became the envisaged World Machine.

The accelerator and detectors were designed to explore the full mass range where the Higgs boson might be found. There are/were two caveats with the Higgs. As a model, it was viewed with some skepticism by the physics community. One of my theory colleagues who was deeply involved with the creation of the unified electroweak model called it the "toilet bowl of the Standard Model."

There was a competing model. The mass of the proton does *not* come from the Higgs mechanism. As far as we know, the up and down quarks are nearly massless, as are the gluons that bind them together. We believe that the dynamics of the quarks and gluons interacting inside the proton are responsible for its mass. A competing model for the mass generation in the Higgs mechanism called *technicolor* was based on the mass generation we see in the proton.

Beyond that was the possibility that we simply wouldn't find the Higgs particle and would be left scratching our head, wondering how nature solved the problems. While designing the detectors for the

LHC, we pondered exploring the W-W scattering in figure 11.17. This appeared to be a challenging task and wasn't an obvious slam-dunk.

After years of data taking, the two main experiments at the LHC, ATLAS and CMS, had statistically significant results to claim a discovery of the Higgs at a mass of 125 GeV, and the announcement was scheduled for a seminar on July 4, 2012. The CERN auditorium, where the seminar was to be held, was large, but not large enough for the anticipated crowd. There was a section up front reserved for VIPs, and the other half of the auditorium was on a first-come first-served basis. Unfortunately, I didn't rise to VIP status, and there was a long queue of mostly summer students lined up on the evening of July 3. I didn't have the stomach to wait in line all night long, so I went to my apartment and then took the tram back to CERN on the Fourth and marched to my office, where I watched a livestream of the seminar on my laptop.[2]

Was that the end of work on the Standard Model? Well, it turns out, no.

The Higgs is a strange creature. To study the Higgs, we must hunt down every aspect of it at the LHC. One consequence of the Higgs is that the interaction strength with other particles is directly related to their masses. The experimental program is to try to test every known coupling with the Higgs to see if the couplings behave this way and it takes a lot more data. Another test is that the Higgs is supposed to interact with itself. This self-interaction gives us a window into possible new physics beyond our Standard Model.

ARE QUARKS AND LEPTONS COMPOSITE?

If you look at the families of leptons (fig. 11.9), it becomes a bit reminiscent of the periodic table of elements yet again. A repeated pattern of particles falling into a regular schema might suggest yet more underlying substructure. Certainly, this suspicion hasn't escaped notice among physicists. In the 1970s and 1980s the concept arose that the quarks and leptons were made up of yet smaller objects called preons.

No one has yet to come up with a viable theory that has quarks and leptons as composite objects. We can search for possible compositeness at the very highest energies the accelerators can deliver. In many ways, accelerators are like ultrapowerful microscopes, allowing us to probe into the smallest distance scales. If quarks or leptons have substructure, there would be additional interactions arising at the highest energies we can probe. Every time a new accelerator comes online or a new energy frontier is probed, the first order of business is to look for signs of compositeness, but thus far, there appears to be no substructure whatsoever, down to 10^{-20} meters.

To give a feel for this 10^{-20} meters, let's put it in human terms. We inhabit a scale of about a meter, roughly speaking. Accelerators act as ultra-powerful microscopes, and our ability to probe down to this scale would be like looking at a virus on Earth from a vantage point on Mars. So, at least at this scale the quarks and leptons seem point-like.

Another clue that there may be no substructure are the masses of the quarks and leptons. In the periodic table of the elements, the repeated regularities of the atomic properties were accompanied by a regular pattern of masses. Likewise, with the Eightfold Way, the regularities of the hadronic properties were accompanied by mass regularities. But, in the case of the leptons and quarks, there is no discernable pattern of masses that accompanies the regularities in their properties. The gap between the heaviest quark, the top quark, and the lightest lepton, the neutrino, is a ratio around 100 billion to one. In contrast, uranium is only 238 times heavier than hydrogen. This would suggest that the quarks and leptons are somehow more primordial, but the pattern of masses remains a mystery.

GRAND UNIFICATION

Recall that the coupling of the strong force gets weaker as energies get higher, and the coupling of the electromagnetic, now the electroweak, gets stronger. If we extrapolate the strength of these couplings, we can find an energy where the two meet. This is suggestive that all three forces: strong, weak, and electromagnetic are manifestations of a single *Grand Unified Theory* (GUT), in the same way the electromagnetic and weak forces combine to make an electro-weak theory.

The GUT energy scale is about 10^{16} GeV, which is 14 orders of magnitude higher in energy than the electroweak unification scale of around 100 GeV that we probe at the LHC. This poses a challenge, although we might imagine all sorts of physics in between the electroweak scale and the GUT scale. Conversely, there might be what theorists call a "desert," meaning that there's no physics between the GUT scale and the electroweak scale. Most physicists believe that there must be a Grand Unified Theory, partly because the charge strengths seem to approach each other at higher energies/shorter distances.

What would a Grand Unified Theory entail? By its very nature it implies that the strong, weak, and electromagnetic forces would be a single force. One prediction of grand unified theories is that protons can decay. One possible decay mode is a positron and a neutral pion. At present, the limits on the proton lifetime indicate that it lives at least 10^{34} years. Considering the universe, as we know it, is about 10^{10} years old, there's not much concern about all the protons decaying.

Experiments with vast quantities of water have been used to detect just a few precious protons decaying. The simplest grand unified theories have been ruled out by the lack of any decay signatures thus far, but a GUT theory remains an ambition of the field.

REMAINING MYSTERIES

Our Standard Model of particle physics works amazingly well in predicting all sorts of experimental results, but it does not exactly appear elegant. There were many details I didn't get into. Just off the top of my head, I count twenty-four free parameters that must be empirically determined, as opposed to originating from first principles. This includes masses and charge strengths. Beyond this, there are "mixings" between quarks and neutrinos. The mixings are beyond the scope of what I've presented here, but these present more free parameters beyond simply the masses. There are also questions beyond the Standard Model, many inspired by astrophysics:

What is dark matter?
What is dark energy?
What drives the asymmetry between matter and anti-matter in the universe?
How are the four forces unified?
What does a quantum theory of gravity look like?

While I could argue that our Standard Model of particle physics and cosmology is the most comprehensive model of the universe yet, these questions and more arise. A major challenge is understanding and even accessing distance scales where we believe all the fundamental forces are unified.

SCALING *to the* MULTIVERSE

We believe that the current model of particle physics is a fundamental theory. Quarks and leptons are indivisible point-like entities and the forces they feel are primordial. There is also a sense that the universe as we know it is becoming better understood. Yet there are many questions remaining and there is a fear that we may be nearing a limit of experimental probes. One can only build accelerators so large. Advances in astronomy require telescopes of such a scale that financing them becomes a challenge.

With unanswered questions, all sorts of theoretical speculations rush in the fill the void. Many of these new ideas are highly imaginative and often capture the public's attention.

SCALING: ANOTHER KIND OF JOURNEY

The history of Western physics and astronomy has been one of ever-expanding horizons toward both the large and the small. When we think of a journey in our mind's eye, we typically imagine moving through an environment with a set scale. I might be on a walk in a park or a drive across Massachusetts. On my walk in the park I am only thinking in terms of the trails, and perhaps how I got there from my house. In a long drive I may be thinking in terms of a network of long-distance highways and roads that crisscross the state.

If I ask you to imagine a space flight through the solar system, you probably picture the planets and not think too much about very distant stars or the details of the Earth. If I ask you to imagine traveling to the center of the galaxy, you might imagine traveling through the spiral arms, passing stellar nebulae. Likewise, if I ask you to picture an atom, you might think about the orbits of electrons around a nucleus like it was a miniature solar system.

All these imaginings are on different scales, but what if we imagined traveling *through* distance scales itself?

Here are some scales to consider:

To the large, we have:

Walkable distance in a day: 16 kilometers—10 miles

Drivable distance in a day: 500 kilometers—300 miles

Radius of the Earth: 6,400 kilometers—4,000 miles (roughly an eighty-day journey, ala Jules Verne)

Earth–Sun distance: 150 million kilometers—100 million miles

Distance to Neptune: 5 billion kilometers, 4 light-hours (4 hours for light from the Sun to reach Neptune)

Distance to nearest star: 4 light-years

Distance to the center of the galaxy: 25,000 light-years

Distance to the nearest spiral galaxy (Andromeda): 2.5 million light-years

Limit of the observable universe: 14 billion light-years

(On this last point, we have to be careful because light that might reach us from this distance has traveled since the time the universe was much younger, but it still sets a scale.)

To the small, we have:

Human stature: 170 centimeters

Human hair: 50 microns (50 millionths of a meter)

Size of a cell: 1 micron

Size of an atom: One 10-billionth of a centimeter

Size of the atomic nucleus: One trillionth of a centimeter
Size of a proton: One-tenth of a trillionth of a centimeter
Smallest distance probed in an accelerator: One-one-hundredth of a
   millionth of a trillionth of a meter (10^{-20} meters)

On the scales that we've been able to probe, there are forty-seven
orders of magnitude (factors of 10) between the largest and the smallest (fig. 12.1).

The location of one meter at the human scale is not an accident;
it was initially defined by the French National Assembly to be one
ten-millionth of the distance from the Equator to the North Pole,
and is quite consistent with the scales of everyday life.

If I were to be egocentric in the manner of our cognitive maps, I
would observe that the center of the above scale of scales is roughly a
kilometer, or the length of a stroll around my neighborhood. I could
argue that this distance scale is privileged among others and centers
on the human. This might be of some personal comfort, but I hasten
to point out that this figure only captures spatial domains that we've
explored and doesn't touch on what may lie beyond the frontiers
of the large and small. The middle of the scale is also something of

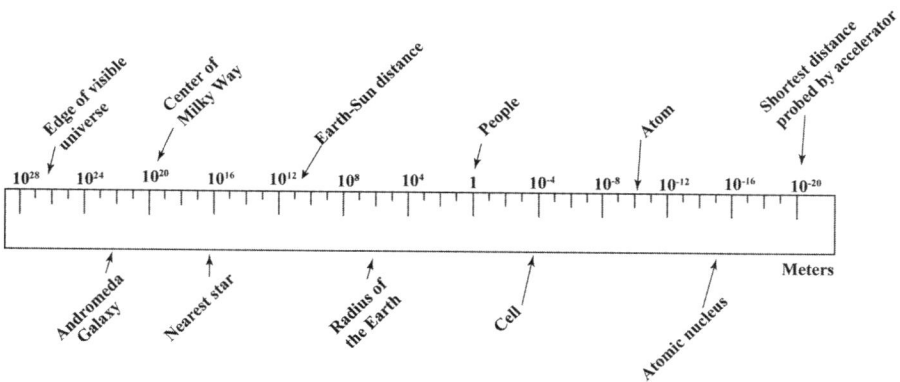

FIGURE 12.1 Distance scales humans have explored
through observation and experiment.

an accident, as the largest and smallest scales are defined somewhat arbitrarily by what we can access by experiment and observation.

A curious feature of our astronomy and physics is that our ability to describe phenomena has taken leaps in the distance scales where the laws of nature seem valid. Accompanying these leaps is also a sense of elegance or beauty in the descriptions. Some examples include:

1. Newton's theory of gravity united the descriptions of motion on Earth and in the sky, providing an accurate model of orbits in the solar system.
2. Hubble's work on galaxies and the realization that we can describe a universe by solutions to Einstein's equations for gravity showed that the universe is far vaster than our Milky Way.
3. Quantum mechanics and particle physics have brought us to a theory that we believe describes fundamental particles: quarks, leptons, and the force carriers.

There is an ambition in the physics community to describe phenomena on all the distance scales with a single unified theory. Physicists, for the most part, believe in a theory that unifies all the fundamental forces. For some time, it is hoped that a Theory of Everything (TOE) would emerge, but it thus far remains elusive.

NATURALNESS AND FINE-TUNING

The Grand Unified Theory (GUT) described in the previous chapter has three of the four fundamental forces (strong, weak, and electromagnetic) converging at a distance scale fourteen orders of magnitude smaller than what we're currently probing with our most powerful accelerators. If our model of particle physics is complete with nothing new, there is a major problem with this large gap called the fine-tuning or the naturalness problem. The fundamental constants of our theory of elementary particles at the scale we're probing seem to be exquisitely balanced to thirty or more decimal places. There must be a cancellation between constants that seems highly unlikely.

To hammer this idea home, let's look at two possible values of pi carried out to thirty digits, with the only difference being in the thirtieth digit past the decimal:

$$3.141592653589793238462643383279$$
$$-3.141592653589793238462643383278$$
$$0.000000000000000000000000000001$$

This accidental subtraction seems so unlikely that physicists believe that there is some new process that intervenes and solves this problem. One concept that's popular is called *supersymmetry*. In supersymmetry, there is a postulated group of particles that mirror our known fundamental particles with a change of spin. The combination of the supersymmetric "partners" with our known fundamental particles creates a cancellation that eliminates the fine-tuning problem. There's one catch: for the supersymmetric partners to do the "job" of eliminating fine-tuning, they must have masses close to the masses of our known particles.

We've been searching for supersymmetric particles for decades and haven't seen anything. It had been speculated that dark matter is composed of supersymmetric particles. If they are found and have the properties we ascribe to dark matter, it would be a major coup. But, as of now, nothing. So far, experimental searches for supersymmetry have come up empty, and we're turning to more exotic scenarios where supersymmetry may be hiding. It's quite possible that there are some new physics processes that are at play in the distance scales out of reach of our accelerators, somewhere between where we are now and the scale of a Grand Unified Theory.

At shorter distances still, there is a destination where we think that all four forces converge, and we believe gravity becomes quantum mechanical. This is known as the Planck scale. By pulling together the constants that would be essential for quantum gravity, we find that it's at 10^{-35} meters. Many of us believe that this is the scale where all four forces are unified. Said perhaps more precisely, there is a Grand Unified Theory at a slightly larger scale that then ties into gravity at this Planck scale. Perhaps the origin of the universe as we know it is tied into this smallest scale we can imagine, but in that case it's challenging to reconcile this with the vast scale of even the visible universe, let alone what might lie beyond.

HIDDEN DIMENSIONS OF SPACE

Another concept to solve the naturalness problem is that the gap of fourteen orders of magnitude of distance scale is not real because there are extra dimensions of space that are somehow hidden or rolled up. This is related to the Kaluza-Klein unified gravity-electromagnetism concept mentioned in the previous chapter. If the extra-dimension theory is the case, then it's possible that the GUT and Planck scales are just around the corner and there no scale gap. One of the predictions of these models is that microscopic black holes could be produced in proton-proton collisions.

This idea of microscopic black holes being created at the Large Hadron Collider (LHC) generated considerable press coverage. The fear was that collisions at the accelerator could produce a black hole that would destroy the Earth. This spawned Internet conspiracy theories about CERN. Big-time news outlets followed the story. *Forbes* ran a piece, "Could the Large Hadron Collider Make an Earth-Killing Black Hole?"[1] *The New York Times* covered it, "Gauging a Collider's Odds of Creating a Black Hole," by science writer Dennis Overbye.[2]

The concept even made it to the courts. There was the case of a woman who sued to keep the LHC from running, leading to an NBC news headline, "German Court Rules That Collider Won't Destroy the Earth." This was in 2012.[3]

There is an interesting personal angle to this. Around this time, out of the blue, I got an e-mail from a stranger in New Mexico. She related how she had a dream about the LHC and the possibility of a black hole destroying the world and that I, John Huth, knew how to prevent this from happening. Perhaps against my better judgment, I engaged her in a discussion about why an LHC-generated black hole won't destroy the world. She turned out not to be a kook, and seemed relatively well grounded, although how my name came up in the context of her dream is a mystery to both of us.

The conventional thinking in the creation of a black hole at the LHC is that it would rapidly evaporate through a process called Hawking radiation, named after theoretical physicist Stephen Hawking. The rapid evaporation would produce huge numbers of high-energy particles that would be visible to the experiments looking at proton-proton collisions, but the black hole would disappear in an instant.

Once, when I was visiting CERN, I bumped into a colleague in the theory group, Michelangelo Mangano, whom I had previously done a lot of work with. We got to talking, and he told me that he was tasked by CERN to examine the question of black-hole production and why or why not it could constitute a threat to the Earth. I mentioned the e-mail from the lady in New Mexico, and how I explained to the woman that they would evaporate immediately from Hawking radiation. Mangano said that

> skepticism of this radiation (which would evaporate any black hole produced at the LHC well before it could even leave the beampipe) was raised by the critics who suggested the LHC was dangerous. Rather than getting into a scientific argument as to whether Hawking radiation exists or not, we felt it was easier to give a proof of LHC safety relying on experimental evidence.[4]

The evidence that Mangano was referring to related to the existence of super-high-energy particles like protons or atomic nuclei flying through outer space. We see these collide in our upper atmosphere, for example. The high-energy cosmic rays reach all manner of astrophysical objects: stars, other planets, and so on. Some collisions between the cosmic rays and stars have the same effective energy that's produced at the LHC. If a stable black hole had been produced, then some astrophysical objects (white dwarf stars) would not have been around as long as they have. So, from the stability of astrophysical objects, we can infer that no long-lived black holes can be produced at the LHC energies. He added that no theory exists that predicts the production and existence of stable microscopic black holes.

STRING THEORY

Although we have a rough idea of the distance scales where quantum effects become important in gravity, we do not yet have a consistent theory of quantum gravity, unlike the other three fundamental forces.

One approach to quantum gravity is string theory. The basic concept is that elementary particles, rather than being points, are strings when considered on a short enough distance scale. The normal diagrams we use to chart out particle interactions show particle interactions as lines that intersect. If fundamental particles were strings, there might be more three-dimensional renditions of the diagrams where we have intersecting surfaces (fig. 12.2).

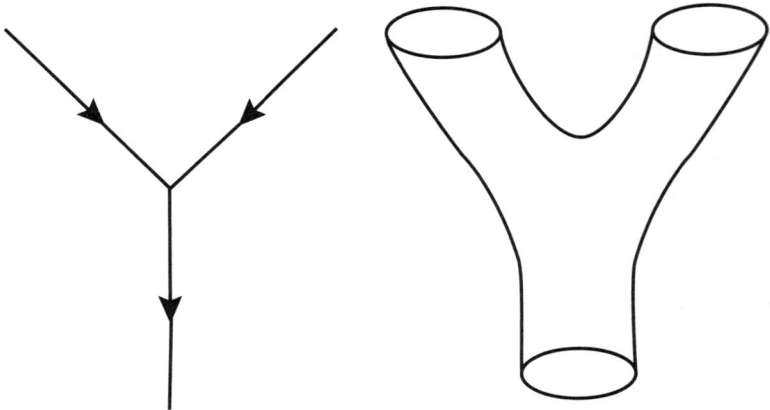

FIGURE 12.2 Feynman diagrams of interactions go from lines intersecting to surfaces.

The different properties of particles we see, like charge, are associated with vibrations of the strings. For finite-sized strings only certain kinds of vibrations are allowed. A guitar string, for example, only produces sounds at certain discrete frequencies. Likewise, the elementary particle strings can have certain kinds of vibrations that are manifested as particle properties.

For quite some time there was hope that string theory would unify particle physics by describing not only quantum gravity, but the other three forces into a theory of everything. In the early studies of strings, the concept seemed highly constrained, and this fact alone made it seem that it would produce a unique answer. String theory exists in ten dimensions (nine spatial, one temporal), and a cousin of string theory, M-Theory, exists in eleven dimensions.

Many aspects of string theory are compelling. For example, in 1997 theorist Juan Maldacena put forth the idea that there was a geometric relationship between gravity and the other fundamental forces. At the time, this seemed like a breakthrough in our understanding of a unity of the forces. String theory has also generated numerous advances in mathematics that might not have been otherwise possible.

This ambition toward a theory of everything is compelling. But the unification scale associated with quantum gravity is so far removed from our everyday reality that it's challenging to come up with experimental tests.

The absence of experimentally verifiable predictions from string theory prompted criticisms. One critic, Peter Woit, published a book in 2007 titled *Not Even Wrong: The Failure of String Theory and the Search for Unity in Physical Law*. The phrase "not even wrong" came from physicist Wolfgang Pauli, who used it to characterize concepts in physics that were not testable and verged on theology. Another critic, Sabine Hossenfelder, published a book in 2018, *Lost in Math: How Beauty Leads Physics Astray*. There is a feeling among some theorists that elegance and beauty should be a guide to physical theories. In some ways this is true, but it's often an after-the-fact "ah-ha" moment when a significant breakthrough happens that reduces a complex group of observations to a single unified principle. Hossenfelder is a popularizer of science with a training in theoretical physics who is a frequent critic of the state of particle physics, particularly theory.

For quite some time, it felt as if string theory were converging on one answer where all the forces could be understood, and even perhaps all the constants of nature. But that hope changed dramatically roughly in the 2000s. Rather than converging on a single theory, developments led to the idea that there may be a huge number of possible solutions—of order 10^{500}. There is speculation of a "string landscape" of many different possible universes with different constants of nature. In the "landscape" there are pockets where a universe might exist with certain constants, but these may be unstable, and the universe could spontaneously "tunnel" into another configuration.

I mentioned quantum tunneling in the chapter on cosmology as a mechanism that could take patches of the universe out of inflation. In terms of the landscape, one might ponder that a solution to a particular universe could tunnel into another universe with different physical constants. At some level, this raises the question of what we mean by the term *universe*. Is it just the visible universe or something more encompassing?

THE MULTIVERSE

This concept of a string landscape is one variant on the idea of a *multiverse*. It has also been invoked to explain the fine-tuning problem, why the one-part-in-10^{30} balancing of physical constants works out.

There are many variations on what constitutes a multiverse. One kind of multiverse is the case of a collection of possible universes that are partly stable but could tunnel into another. In some cases, a new universe could bud out of an older one as a patch of space spontaneously tunnels into another configuration. All of these budding universes could have different constants of nature and would look very different from our universe.

Along with this is the idea that when our early universe underwent the inflation phase, many patches of space tunneled into different semi-stable universes but were surrounded by still rapidly expanding/inflating space. Physicist Brian Greene made an analogy to the structure of Swiss cheese, where the holes in the Swiss cheese are the universes, and the cheese itself is a kind of scaffolding that is expanding so rapidly that the universes (the holes) don't have a chance to communicate with each other. The universes (holes) may be expanding on their own, or perhaps collapsing, but the cheese itself is expanding even faster. But herein lies a problem. If there are other universes out there, but we cannot connect with them, we cannot really test the proposition.

As if that's not enough, there's an even more far-out version associated with the Many-Worlds Interpretation of quantum mechanics. If you recall from the chapter on quantum mechanics, there is this curious problem of wave-packet collapse. When a particle is a wave, it can be spread out in space. But when we try to make a measurement of it, or it interacts with something, suddenly it gets localized to a single point in space. It seems as if all that information of the wave packet is suddenly lost as it's now in one location. The Many-Worlds Interpretation seeks to preserve that information by positing that when a wave-packet collapse happens, it happens in many universes where the different possible outcomes are realized. A more recent twist on this was to look at this as real universes being spawned whenever a collapse happens.

This Many-Worlds multiverse had proven to be a huge cultural phenomenon. Writ broadly, it suggests the metaphor that for every decision, every interaction, there are multiple versions of yourself and others who play out the possibilities. If you break up with a significant other, there's another universe where you stick together. If you barely miss getting hit by a truck by not stepping into a crosswalk too soon, there's another universe where you get hit by that truck.

Since roughly 2011 onward, movies, plays, TV series, novels, even comic books have employed the multiverse concept as foundational plot devices. When I first brought the topic of this book to a seminar for First-Year Harvard students in 2017, one kept asking me, "When are we going to discuss parallel universes?" Being a culturally out-of-touch physicist, I didn't know why he was so keen on learning about parallel universes. I credit the student with opening my eyes to this as a popular phenomenon, and I immediately saw how it would appeal.

In a *New Scientist* piece, author Rowan Hooper waxes eloquent about the possibilities,

> I'm rich. I'm a movie star. I'm king of the world. I'm also poor. I'm homeless. Lots of me are dead.
>
> I'm none of these. Not in this universe. But in the multiverse I'm all of them, and more.[5]

As with other overwrought cultural phenomena, there was some backlash as it got a little old for certain observers. In an article in *The Atlantic*, Sam Kris laments

> The multiverse is a prop, a way to explain away things that can't otherwise be explained. It's supposed to induce a Copernican vertigo, your own tininess in a hall of mirrors where every reflection can strut around asserting its primacy, but in fact, it's a strangely comforting doctrine.[6]

Returning to the fine-tuning problem and the question of scale, with our current science it seems that we live in a highly improbable universe. If we compare the size of the universe to the distance scale where we believe forces may be fully unified, we may end up with a hundred or more orders of magnitude of scale to be explained.

We may just happen to live in a universe that supports life, however improbable. That is to say, there exists simultaneously a huge number of universes with different physical constants and we just happen to inhabit one of them. We ended up in one of the 10^{500} possibilities that supports life.

I raised the question of the anthropic principle in the chapter on cosmology. This has also been invoked to explain how we just happen to be in a universe where an improbable combination of constants support life.

Many physicists view the anthropic principle as a kind of capitulation. Another problem is one of observability. If some distant part of the universe tunneled into a different universe with different constants of nature, how would we even know about it? It also raises the question whether this is still science if it can't be tested.

Nonetheless, the desire to have a robust explanation of our universe from first principles is compelling to many physicists. Others, perhaps a minority, have started to embrace the anthropic principle. Jim Peebles, one of the physicists who predicted the cosmic background radiation, recently wrote a paper titled "The Physicists Philosophy of Physics," where he ponders a more sociological/philosophical view of how the community of physicists may or may not converge on a worldview.[7] He sees the anthropic principle connected to the huge scale between the Planck and the observable universe as generating a schism. Recall the cosmological constant Lambda, Λ, associated with the acceleration of the expansion of the universe. It is tiny compared with what we think might be the scale associated with the constant necessary to generate inflation in the early universe or that expected from quantum gravity. Of this, Peebles writes:

A multiverse of universes is expected in some versions of the cosmological inflation picture of what our universe was doing in the very early stages of expansion. It would be interesting to compare the value of Λ derived from cosmology to the range of values expected in those universes in a multiverse that have physics capable of supporting life of a kind that would take an interest in the value of Λ. But we do not have an adequate theory of the properties of universes in a multiverse, or of the kinds of physical theories that allow life. If research continues to fail to reconcile the value of Λ from quantum physics with the value from cosmology, we can anticipate two physicist camps. One would insist on working even harder to avoid the anthropic argument. The other would accept a multiverse. The latter risks missing discovery of a perfectly good resolution by a sensible improvement of standard physics. The former risks spending a lot of time looking for that improvement with no guarantee of success. You choose.[8]

Here, I contrast two physicists who represent the camps Peebles allude to. Stanford physicist Leonard Susskind married the concept of string landscapes to multiverses and is something of a proponent of the anthropic principle. In a 2020 interview, when asked to reflect on the views of physicists toward the anthropic principle, he said:

You have to go back a dozen years or so to what physicists thought about these anthropic ideas. They thought of them as being religious. They thought of them as being the statement "the world is the way it is because God made it so that people could exist."

Physicists don't like to explain things that way. They want hard-nosed equations. They want hard-nosed explanations based on the laws of science and so forth. So they were very very, uh, I would almost say angry. There was almost an anger at this idea that the universe was somehow made for us to live in.

That mindset was the mindset about the anthropic principle, say, ten, twelve years ago. It just rankled against physicists' objectivity to introduce something as personal as saying the world was made for us to exist in and live in. But, as time went on, they found it less and less possible to explain the world as we see it, and new kinds of explanations evolved, which were not religious in the least. They depended on this enormous diversity, but they had in the back of their mind always that there was some supernatural aspect to it.

The other thing was ambition. I don't mean personal ambition, I mean the ambition of the field. The ambition of the field was to explain every single fact of nature; the constants of nature, the particles of nature. The hope was that we would be able to use our equations to explain every single feature of nature, and that just conflicted with the idea that nature could be enormously diverse.[9]

As a counterpoint, physicist Paul Steinhardt views the multiverse as a kind of nonscientific abdication:

> Just imagine—your theory was supposed to explain why things are this way and not that way, or more probably this way than that way. Instead, what you find is—your theory predicts that all are equally possible. So some of the proponents would say, "Great, no experiment can possibly disprove our theory now. We can learn about our patch of the multiverse, our particular environment, but we can never disprove this idea." And I would say, "Well, this is disaster, because your theory now fails the fundamental criterion of what makes a theory scientific—it's now unfalsifiable—impervious to any experimental test." Now, all this has been clouded by the way people talk about the subject.[10]

Using the anthropic principle to explain what seems like a highly improbable universe existing in a vast multiverse has some kinship to the egocentric cognitive map. In this formulation, our universe is referenced to humans. The "ambition" that Susskind speaks of is a more allocentric viewpoint that suggests that we're in a unique universe that doesn't need to reference humans. Although a theory has a huge number of possible universes, this doesn't mean it's necessarily correct nor that there isn't some other theory or additional organizing principle that selects our universe.

While it isn't derived from first principles in physics, there is a model by cosmologists Neil Turok and Latham Boyle where they added some unique properties and invoked the principle of entropy to come up with a universe like ours.[11]

Entropy is a thermodynamic principle that's related to the probability of finding a system with many possible states in a particular set of states. If we have a jar full of air in a room that's otherwise evacuated of all air and then open the jar, the air molecules will flood the room.

In principle, there is nothing that rules out the possibility that the air molecules will later spontaneously converge back into the jar. But the number of possible states of the air molecules permeating the room is vastly larger, and much more probable than the state where the air molecules return to the jar. This is an example of entropy.

By incorporating entropy with the more standard ingredients of cosmology, Turok and Boyle demonstrated a model that suggests that universes like ours are the most probable. Now, the entropy they added was somewhat ad hoc, but if you assume it, it explains the age and flatness of our universe as the most probable state. Turok is quite candid about the ad hoc nature of the addition of entropy, "What we've used is a cheap trick to get the answer without knowing what the theory is."[12]

Curiously, there is another entropy principle associated with the mass of the Higgs boson.[13] In this model there is also an ad hoc introduction of entropy as an organizing principle associated with the strength of the coupling to the particles it decays into. If one assumes this principle, the Higgs mass is predicted. Perhaps "post-dicted" is a better way of phrasing it, as the model was published when the Higgs mass was already known. I'm not claiming that these two models are correct or that entropy is an important new organizing principle, but it is a kind of "existence proof," that there are other possible organizing principles out there that don't rely on the anthropic.

Imagination often outruns our ability to test. *War of the Worlds* was published in the 1890s and the first probes to Mars were launched in the 1960s. The Higgs boson was proposed in 1964 and discovered in 2012. There are presently accelerators on the drawing board that can probe aspects of the fine-tuning problem. The James Webb Space Telescope has given us an amazing window into the early universe and there are powerful ground-based telescopes coming online and in the planning phases. Historically there have been startling breakthroughs, although the time between breakthroughs makes one feel that there are long plateaus where time is spent plugging away until the next major jump in knowledge occurs.

THE PSYCHOLOGY *of*
SPACE FLIGHT

Until this chapter, I've engaged in visions of space imagined by the earthbound. With powerful accelerators and telescopes, we've managed to explore the inner space of subatomic particles and outer space nearly to the dawn of time. It's been remarkable that we gotten this far without leaving the planet. But humans are travelers. We found our way out of Africa in successive waves of migration and somehow figured out how to live in nearly every environment on the planet. What about physically leaving the Earth?

The ambition is not entirely new. Recall the myth of Daedalus and Icarus. Daedalus was the craftsman who fashioned the Labyrinth to imprison the Minotaur for King Minos of Crete. As per the mythology, Minos consigned his enemies to the Labyrinth where they would be devoured by the Minotaur. The relationship between Minos and Daedalus fell apart when Daedalus enlisted Princess Ariadne to give Theseus the magic thread to escape the maze and the Minotaur. Furious, Minos imprisoned Daedalus and his son Icarus in the Labyrinth.

Clever Daedalus realized that his imprisonment was by virtue of being tethered to the two-dimensionality of the Labyrinth, but there was the additional vertical dimension that he could access. Accordingly, he fashioned wings out of osier branches and wax so that he and Icarus could escape by flying upwards.

Daedalus warned Icarus to be cautious and stay away from the heat of the Sun, but Icarus was exhilarated by his newly developed ability to fly, swooping low to the ocean and then high to the sky. But, as we know from the legend, Icarus flew too close to the Sun, where wax began to melt and then fell to the sea and drowned. The myth has long been seen as a story of both human cleverness and hubris.

The modern era of spaceflight also had its origins in the human imagination. Many of the early rocket pioneers were inspired by Jules Verne's novel *From the Earth to the Moon*, published in 1865. The story line is that a group of weapons devotees, The Baltimore Gun Club, set out to build a giant cannon named *Columbiad* that could fire a large projectile with passengers to the Moon. Three early pioneers of rocketry: German Hermann Oberth, Russian Konstantin Tsiolkovsky, and Peruvian Pedro Paulet all reported that they were directly inspired by Verne's novel.

Science fiction also inspired Robert Goddard, the American pioneer of rockets, but in his case, it was by H. G. Wells' *War of the Worlds*. Wells imagines that the propulsion of the Martians to Earth may have been via a cannon. In the opening chapter he describes a telegram reporting observations of flashes of light from Mars from an astronomer:

> This jet of fire had become visible about a quarter past twelve. He [the astronomer] compared it to a colossal puff of flame suddenly and violently squirted out of the planet, "as flaming gases rushed out of a gun."

In addition to Verne, another influence on the rocket pioneer Tsiolkovsky was a cultural movement at the turn of the nineteenth into the twentieth century called *Russian Cosmism* that prophesized a kind of manifest destiny for humans to populate outer space. He was also a proponent of a concept called *panpsychism*, which is the idea that the universe/reality has a fundamental mind-like component to it. Along the lines of the wisdom of the inward parts, Tsiolkovsky felt that atoms themselves had intellect. He also expressed the belief that intelligent life should be common throughout the universe and believed that the Milky Way would ultimately be colonized by humans.

All four rocket innovators—Goddard, Oberth, Paulet, and Tsiolkovsky—dreamed of being able to travel to the Moon and planets through outer space. Their discoveries were mostly independent of each other, including making the necessary calculations to understand the speeds required to reach Earth's orbit, and then escape velocity. They all realized that the acceleration from a cannon was far too great to be feasible, as the humans could not withstand the high g-forces. The pioneers came up with the common solution of rockets, where the thrust of burning propellants out the back of the rocket could provide a more controlled acceleration that would allow humans to be carried up into space.

The concept of rockets was well established, as Chinese fighters had used them as weapons since the thirteenth century. The idea of using rockets to launch humans into space, however, was new. The oldest military rockets used solid propellants like gunpowder. These, however, were not as controllable or efficient as liquid propellants, which were the focus of the early pioneers.

One major obstacle to rocket flight is the need for both a fuel and an oxidizer to burn the fuel in the vacuum of space. In Goddard's early 1926 test, he used gasoline and liquid oxygen as the fuel/oxidizer combination for the very first liquid-fueled rocket.

Another approach for a fuel/oxidizer combination was a hypergolic propellant, where both the fuel and oxidizer could be held at room temperature. When in contact, the two chemicals would combust and provide thrust. Pedro Paulet experimented with gasoline and nitrogen peroxide as an early attempt at hypergolic propulsion. The hypergolic combinations have the downside of being very reactive.

Some of the earliest rockets in regular use like the Vostok that propelled Yuri Gagarin into space, were combinations of liquid oxygen and propellants like alcohol or kerosene-derived chemicals. The Titan rockets used later in the US Gemini program were adopted from ballistic missile rockets that employed room-temperature hypergolic combinations, as the missiles had to be launched on a moment's notice from silos in the event of a nuclear exchange with the Soviet Union.

Two other major technological hurdles were necessary for feasible rockets: fuel pumps and guidance systems. For the propellant to reach the rocket nozzles at a sufficient rate, pumps had to be designed to bring the fuel and oxidizer to the nozzles. For the German V2 weapons developed in World War II, this was accomplished with steam-powered turbopumps. The turbopumps themselves were fired by a mixture of hydrogen peroxide and potassium permanganate, which produced steam, spinning the turbines of the pumps.

The question of stabilization and direction of flight was another problem to be solved. Gyroscopes and devices called accelerometers could be used to sense the rocket's orientation and motion. By producing a feedback mechanism between these sensors and fins in the combustion chamber, the thrust of the rocket could be directed in a way to achieve stable flight. This became known as inertial guidance. These systems did not always work. Spectacular film footage of rockets where this system has gone awry shows them following an erratic wobbly path until they crash.

Robert Goddard pursued solutions to these challenges with increasing sophistication. Unable to convince the US government to fund his efforts, he relied on private funding from Daniel Guggenheim, and the Smithsonian. When Goddard published his ideas on rocketry early on as a Smithsonian bulletin it was met with criticism, even a touch of ridicule. Perhaps the most famous of these was a *New York Times* editorial that scoffed at the concept:

> That Professor Goddard, with his "chair" in Clark College and the countenancing of the Smithsonian Institution, does not know the relation of action to reaction, and of the need to have something better than a vacuum against which to react—to say that would be absurd. Of course he only seems to lack the knowledge ladled out daily in high schools.[1]

Perhaps it doesn't bear comment on my part, but an important aspect of Newton's laws (as we saw in chapter 7) was the concept "for every action, there is an equal and opposite reaction." The rapid expulsion of the burning propellant is the very principle that rockets need to accelerate.

The *New York Times* opinion notwithstanding, the idea of rocket travel in outer space soon made its way into US popular culture. Perhaps most notable was the comic strip *Buck Rogers*, beginning in 1929. Buck Rogers movie shorts were also a staple into the late 1930s and beyond.

Although Goddard imagined the modern uses of rockets to launch satellites and humans, the US government did not take seriously the possibility of rockets in the capacities Goddard imagined. It was only after the Germans successfully developed the V2 rockets employed in World War II that the United States and Soviet Union took serious notice because they could deliver a warhead a long distance over a very short period of time. The one application that the US Army considered seriously at that point was the use of rockets to help boost the launching of aircraft in what was known as Jet Assisted Take Off (JATO).

The development of rockets in Germany between wars had a significantly different evolution. Hermann Oberth was originally from Transylvania and moved to Germany. From a young age, he was transfixed by the concept of rocket travel, inspired by Verne's novels. In Germany he founded a club called Verein fur Raumschiffahrt (Society for Space Travel) in 1927. While Goddard was something of a loner, the VfR attracted many rocket enthusiasts, including the young Wernher von Braun, who would later lead the US space program after World War II.

As with the United States, the concept of rocket flight entered popular culture in Germany. In 1929 director Fritz Lang released the movie *Frau im Mond* (Woman on the Moon), which prominently features rocket travel to the far side of the Moon. *Frau im Mond* is often cited as the first instance of a countdown to zero before liftoff. Oberth was an advisor on the film and cared deeply about the technical details of the rocketry portrayed in the movie. Today countdowns are not merely a dramatic device but serve the purpose of synchronizing the complex operations necessary for the rocket ignition and lift-off.

Von Braun had read Oberth's book *Die Rakete zu den Planetenräumen* (The rocket into interplanetary spaces) when he was younger, and apprenticed himself to Oberth. Von Braun wrote his doctoral dissertation on the practicalities of a liquid-fueled rocket, which was completed in 1934. While the United States seemed uninterested on Goddard's work, von Braun's efforts attracted the attention of the German military who helped fund it. A German artillery captain, Walter Dornberger, developed an interest in von Braun's work and was ultimately tapped to lead the German *Vergeltungswaffen* (Revenge weapon) program in Peenemünde on the Baltic Sea. The V weapons (V1 and V2) were conceived as revenge weapons for the Allied bombing of German cities. The V1 was an unpiloted airplane powered by a jet engine, known as a "buzz bomb" for the sound they made. The V2 was the true liquid-fueled rocket that the pioneers like Goddard envisioned. It flew faster than sound, so it was not heard as it landed in London before exploding.

The facility at Peenemünde represented the first industrial-scale production of missiles and included forced labor associated with concentration camps. The British learned of the facility and in 1943 carpet-bombed the region that they believed housed the scientists and engineers. Subsequently, the production of V2s was shifted to an underground facility near Nordhausen in Thuringia.

The V2 did not have a significant impact on World War II, but its use attracted the attention of both the United States and the Soviet Union. This, along with the development of nuclear weapons, alerted both countries to a combination of weapons that could be quite powerful: an intercontinental ballistic missile that carried nuclear warheads.

Toward the end of World War II, Dornberger and von Braun agreed to surrender as many engineers and documents as they could to the United States at the end of the war. The Soviet Army was closing in on Peenemünde. The entourage traveled close to the US front lines and ultimately surrendered.

The German rocket team was transferred to the United States and initially built a set of V2 rockets for testing purposes.[2] For space flight, it was recognized that rockets had to have multiple stages to reach Earth's orbit and beyond. The V2s did not have staging and could only reach London during the war. In 1948 a combination of V2 and US technology produced the first multistaged rocket. The US program started out in Texas and testing was done at the White Sands Proving Grounds in New Mexico.

In 1950 the rocket team under von Braun were transferred to Huntsville, Alabama, where work began in earnest on the Redstone rocket program, designed to create nuclear-tipped ballistic missiles. Both the United States and the Soviets recognized the propaganda value of launching humans into space, so a dual purpose of weaponry and space exploration became the goals for programs in both countries. Von Braun dusted off his original interest in rocketry for putting humans in space and began to advocate for manned spaceflight.

The Redstone rocket series, like the V2s, used liquid propellants but used multiple stages. For civilian purposes the Redstones were also adapted for placing satellites in orbit, and humans in the Mercury astronaut program. The Soviet Union likewise developed multistage liquid-propelled rockets designated as Sputnik and Vostok.

Historically, the start of the space race is marked by the launch of the Sputnik 1 satellite on October 4, 1957. It was not widely thought that the Soviets had the capabilities of putting objects into orbit and this triggered the "Sputnik crisis." This resulted in an era where spending on scientific education and research increased for quite some time. On November 13, 1957, President Eisenhower delivered a speech where he lamented that the United States had fallen behind the Soviets on scientific training. He committed the United States to increasing the number of US scientists.

This trend is disturbing. Indeed, according to my scientific advisers, this is for the American people the most critical problem of all. My scientific advisers place this problem above all other immediate tasks of producing missiles, of developing new techniques in the Armed Services. We need scientists in the 10 years ahead. They say we need them by thousands more than we're now presently planning to have.[3]

I personally recall the emphasis on science education in the 1960s, and the use of space flight to attract youngsters into the sciences. On trips to the Philadelphia science museum, the Franklin Institute, one of my favorite things to do was sit in the mock-up of a Gemini space capsule and flip the (nonfunctional) switches.

For a time, there was a back-and-forth between the Soviet Union and the United States in spaceflight. Almost in response to Sputnik, the United States launched Explorer 1 on January 31, 1958. NASA was created on April 2, 1958. Soviet cosmonaut Yuri Gagarin became the first person in space on April 12, 1961, soon followed by American Alan Shepard on May 5, 1961. Valentina Tereshkova became the first woman in space in 1963. The first human-made object to reach the lunar surface was the Soviet Luna 2 in 1959.

In this intense period, the Soviets were racking up firsts all around—first satellite, the first man in space, the first object on the moon, the first woman in space, and first space-walk. President Kennedy became convinced of the need to demonstrate Cold War superiority over the Soviets with a stunning achievement in space, and announced to Congress in 1961 that the United States

should commit itself to achieving the goal, before this decade is out, of landing a man on the Moon and returning him safely to the Earth.[4]

HIGH-ALTITUDE EXPERIENCE AND ISOLATION

With the coming of the space age, it was natural to wonder how the real-life experience of traveling high into the sky and into outer space would affect pilots. After World War II and into the 1960s the US Air

Force launched a program of experimental aircraft, designated by the letter X. Chuck Yeager was the first person to break the sound barrier in the X-1. The X-15 jet airplane holds the record for the highest and fastest fixed-wing aircraft, with an altitude of 354,200 feet and a speed of 4,520 mph (Mach 6.7). During this period the use of jet aircraft in military aviation was steadily on the rise.

With a new emphasis on putting a human in space, the question naturally arose about how people would respond to an unforgivingly hostile environment. There were also questions about the limits to conditions that could be endured. An Air Force colonel, John Stapp, experimented with decelerations in rocket sleds and found that people could endure accelerations beyond 30 gs.

The Air Force employed physicians as flight surgeons to certify the fitness of pilots to fly. In addition to physical requirements, there were also psychological requirements, but there was a need to study this. One major concern was the effects of isolation, particularly at high altitude. Although flight at times can produce an exhilaration, a solo pilot at high altitudes or a solo astronaut could be quite isolated and engaged in monotonous tasks.

In popular culture there emerged the image of a "steely eyed missile man" in the jet pilots and rocket engineers of that era. But the truth of the psychology of the pilots during that era was more complex. In 1956 jet pilots related a feeling that they labeled "break-off" as "a feeling of physical separation from the Earth when piloting an aircraft at high altitude."[5]

It was recognized back then that there would be a reluctance for pilots to report psychological problems experienced while flying. For one, it goes against the well-established image of pilots as being tough and stoic. This stoic image was famously documented in Tom Wolfe's *The Right Stuff*, a book about the Mercury era astronauts. There is also a more obvious reason for hesitancy in reporting negative experiences in flight: getting grounded or not selected for future missions.

The initial studies of break-off phenomena were spearheaded by Brant Clark, a psychology professor at San Jose State College, and Ashton Graybiel from the Naval Aeromedical Institute in Pensacola,

Florida. The findings were published in 1957 and were based on interviews with 137 aviators.[6] Clark and Graybiel found that 47 of the pilots reported feelings consistent with the break-off phenomenon. "Those pilots who experienced it characterized the break-off effect as a feeling of being isolated, detached, or separated physically from the Earth. They perceived themselves as somehow losing their connection with the world."

One pilot reported, "You do have a feeling of loneliness . . . It's very lonely alone at high-altitude. I'd rather fly at 20,000 or 25,000 feet. . . . At 44,000 feet you are pretty lonely up there." On the other hand, some pilots expressed a sense of elation, "It seems so peaceful; it seems like you are in another world. . . . I feel like I have broken the bonds of the terrestrial sphere." Clark and Graybiel concluded that it was a real effect: "a clearly defined . . . condition of spatial orientation in which the pilot conceives himself to be isolated, detached, and physically separated from the earth."

Project Manhigh started in 1955 to test the effects of ultra-high altitude on subjects prior to the manned space program. Among the many goals was to assess the effects of cosmic rays on humans high above the shielding effect of Earth's atmosphere. The subjects were lifted in a balloon high into the stratosphere and occupied a small gondola/capsule that held pressure. The gondola/capsule (fig 13.1) was the size of a telephone booth, and the occupants could not stand up. Major David Simmons was a physician in the Air Force. He rode Manhigh II up to an altitude of over 100,000 feet (30 km) for a duration of twenty-four hours in 1957. For reference, modern commercial jetliners fly between 30,000 to 40,000 feet. Before embarking, he managed to take a twenty-four-hour claustrophobia test in the gondola/capsule without issues. But when he was in the air on the flight, he said that he had the experience of break-off:

I experienced a sense of detachment from the Earth at four different times. The first was before sunset when the cloud formations gave a "cliff" effect which provided a frame of reference that helped to emphasize the true vertical distances involved . . . during the night I felt in much closer contact with the stars and space above than I did with the beautiful, but remote clouds below.[7]

The effects of isolation intensified toward the end of the Manhigh II flight. Simons reports:

During the last five hours of the run I had a steady stream of illusions and hallucinations: a helmeted soldier's face staring at me; a feeling that there were people perched around me; a feeling that all my surroundings except the instrument panel were as squiggly as a poorly tuned TV set.[8]

Simons expressed concern that aviators and astronauts could experience a kind of temporary hypnosis that could impair their ability to carry out a mission. He compared this state of consciousness to reports from subjects in sensory deprivation tanks in tests carried out by psychologist John Lilly, who developed a sensory deprivation tank in the 1950s. It contains water held at body temperature with a high concentration of Epsom salts, allowing the body to float on the surface. The tank is dark and there is no sound. Within three hours in the tank, subjects would experience hallucinations.

Even without Lilly's tank, isolation could be simulated more readily with similar results. Lieutenant Colonel Martin Giffen reports on one such study:

Another experimental effort consisted in having subjects lie in a bed with frosted glasses covering their eyes. After several hours, directed and organized thinking became progressively more difficult, suggestibility was greatly increased, and the need for extrinsic sensory stimuli and body motion became more acute. Most subjects found they could not tolerate the experiment for more than 72 hours. Those who remained longer than 72 hours usually developed overt hallucinatory experiences and delusions, similar to those reported from mescaline and lysergic acid diethylamide (LSD).[9]

As a side note, Lilly was interested in mind control, something the CIA was quite keen on in the late 1950s and early 1960s, The use of isolation and LSD were studied in mind-control experiments. Lilly went on to take LSD himself in isolation tanks, and believed he could communicate with dolphins.

While the isolation tanks gave one clue to the origins of break-off, the Air Force, and later NASA developed flight simulators, where the capsule and controls and timing would mimic the psychological experience of being solo in flight and space.

NASA psychiatrist Bryce Hartman reported on the results of the simulated solo missions:

> Several reported that the dials on the task panel sometimes looked like faces or other figures. Other subjects heard music or voices, had the sensation that their arms and legs were enlarged, felt that someone was sitting beside them in the chamber . . .[10]

On this last point, and as the name might imply, a person who experiences a felt presence has the sense that another is with them even though they are alone. A famous example of a felt presence was the experience of three members of Ernest Shackleton's Antarctic expedition when they crossed over the challenging interior of South Georgia Island to seek help: navigator Frank Worsley, seaman Tom Crean, and Shackleton himself. It was an extraordinary effort by the explorers at the end of their rope. All three later said that they felt the presence of a fourth walking beside them. T. S. Eliot was inspired by this report to include this in his poem *The Waste Land*:

> Who is the third who walks always beside you?
> When I count, there are only you and I together.
> But when I look ahead up the white road,
> there is always another one walking beside you[11]

FIGURE 13.1 The capsule/gondola for Project Manhigh. *Source*: National Museum of the US Air Force, Wright-Patterson Air Force Base.

Eliot only vaguely recalled the report from the Shackleton expedition, and had two, rather than three real companions in this snippet.

NASA EFFORTS

NASA was founded on July 29, 1958, when President Eisenhower signed the National Aeronautic and Space Act. It was explicitly made a civilian enterprise, and, as such, was something of a break from the Air Force, which had previously carried out studies on the physical and psychological effects of high altitudes and extreme conditions.

The first focus of NASA was Project Mercury: using rockets originally designed for missiles, the Redstone and Atlas, but instead of a warhead on top, there was a capsule for a single astronaut. The first two missions were suborbital and the others orbited the Earth. Initially NASA had a Space Task Force Group (STG) to determine who might fly. The STG was looking for nonmilitary candidates, including deep-sea divers, mountaineers, flight surgeons, and scientists. In addition, commercial pilots and military test pilots were under consideration.

Eisenhower personally overruled the STG and insisted that only military test pilots could be candidates for the Mercury program. Qualifications included:

1. Age between 25 and 40
2. Height less than 5′11″ (to fit into capsule)
3. Excellent physical condition
4. Bachelor's degree or equivalent
5. Graduate of test pilot school
6. Minimum of 1,500 hours of flying time
7. Qualified jet pilot

It's not 100 percent clear why Eisenhower overruled the STG, but his decision simplified the acceptance criteria. Additionally, the kinds of tasks for a Mercury astronaut were quite consistent with the expectations for a test pilot: isolation, experience with high altitude, routine, monotonous tasks performed in a challenging environment.

Perhaps unspoken in the qualifications: all candidates were white males. There was a consideration of women being part of the Mercury program. In fact, at the start of the Mercury program, there was a Women in Space Earliest (WISE) program. However, while the Soviets launched Valentina Tereshkova in 1963, it wasn't until 1983 when the first US woman, Sally Ride, took the space shuttle into orbit. There was early training of women candidates in WISE, but this was ended as it was felt that it was a needless diversion of resources. In *Right Stuff, Wrong Sex*, Margaret Weitekamp writes:

> At a very basic level, it never occurred to American decision makers to seriously consider a woman astronaut. In the late 1950s and early 1960s, NASA officials and other American space policy makers remained unconscious of the way their calculations incorporated postwar beliefs about men's and women's roles.
>
> Within the civilian space agency, the macho ethos of test piloting and military aviation remained intact. The tacit acceptance that military jet test pilots sometimes drank too much (and often drove too fast) complemented the expectation that women wore gloves and high heels-and did not fly spaceships.[12]

From the Mercury selection criteria, 110 volunteers were initially chosen. After a first pass, this was winnowed down to 32 candidates who were then put through a rigorous testing program. This included psychological testing by two psychiatrists, George Ruff and Edwin Levy. In addition to rigorous physical tests, there were interviews with the psychiatrists and standardized psychometric tests. The physical stresses were already understood: launch and reentry subjected the astronauts to intense vibrations and g-forces. In orbit, there was weightlessness and isolation. The psychological stresses could only be inferred from the experiences of test pilots, simulations, and experiences such as Simons' on Manhigh II.

NASA's website on the selection process for the Mercury program describes the testing process. Here is commentary on a phase of testing at Wright-Patterson Air Force Base in Dayton, Ohio:

> Continuous psychiatric interviews, the necessity of living with two psychologists throughout the week, and extensive self-examination through a battery of 13 psychological tests for personality and motivation, and another dozen different tests on intellectual functions and special aptitudes-these were all part of the week of truth at Dayton. Two of the more interesting personality and motivation studies seemed like parlor games at first, until it became evident how profound an exercise in Socratic introspection was implied by conscientious answers to the test questions "Who am I?" and "Whom would you assign to the mission if you could not go yourself?" In the first case, by requiring the subject to write down 20 definitional identifications of himself, ranked in order of significance, and interpreted projectively, the psychologists elicited information on identity and perception of social roles. In the peer ratings, each candidate was asked which of the other members of the group of five accompanying him through this phase of the program he liked best, which one he would like to accompany him on a two-man mission, and whom he would substitute for himself. Candidates who had proceeded this far in the selection process all agreed with one who complained, "Nothing is sacred any more."[13]

There were two broad psychological characteristics that were looked at in qualifications for astronauts: abilities and personality.

> In terms of aptitude and ability, they include high intelligence, general scientific knowledge and research skills, a good understanding of engineering, knowledge of operational procedures for aircraft and missiles, and psychomotor skills such as those used to operate aircraft. As regards personality astronauts were to demonstrate a strong motivation to participate in the program, high tolerance for stress, good decision-making skills, emotional maturity, and the ability to work with others.[14]

Once the Mercury flights commenced, it might seem obvious from a scientific perspective that data on the astronauts in flight and postflight would be valuable to correlate with the preflight testing program. However, in 1962, the psychology program was abruptly terminated by NASA's new medical director, Charles Berry. All the original data were taken away and vanished.[15]

Fast forward to 1984 and the shuttle program, when NASA hired psychiatrist Patricia Santy to work at the Johnson Space Center. She became interested in the early part of the space program: Mercury, Gemini, and Apollo, and searched in vain for any information about the psychological data from that era. Dr. Santy made a presentation to an outside review panel charged with scrutinizing the astronaut selection program. She was asked to present a history of the psychological studies and said that the data were not available for the Mercury program. She'd looked. The former director of the Manned Spacecraft Center, Chris Kraft, chaired the panel. In her talk to the panel, Dr. Santy said that there was no data available from that era. According to Santy, Kraft became disturbed and interrupted her, "Young lady, you are a dangerous person and are out to destroy NASA! I will not permit that to happen." She again stated that she had tried to find the data and couldn't. Again, Kraft, "You never will because it's [the data] in here [pointing to his head], and it's going to stay there so that people like you can't use it against NASA."[16]

In *Choosing the Right Stuff*, Santy's book on the psychology and the selection of astronauts, she speculates that NASA may have had something to hide. From the isolation studies in the 1950s, Simons's experience in Manhigh II, and the experience with break-off phenomenon, it's tempting to wonder whether some of the astronauts in the single-person Mercury program may have experienced similar symptoms. Why hide it? Santy speculates,

> NASA has always had an intense political need to publicly present the astronauts as national heroes and therefore without psychological flaws or weaknesses. Even the psychiatrists were willing to admit that the astronauts were in many ways remarkable men.[17]

While the psychological data may be lacking, we do have first-hand accounts from astronauts in the Mercury, Gemini, and Apollo programs. One in-depth look comes from Michael Collins, who piloted the command module for Apollo 11, which carried crew members Neil Armstrong and Buzz Aldrin to the first lunar landing.

While Armstrong and Aldrin were walking on the lunar surface, Collins orbited overhead, alone. One could make the case that Collins may have been the most isolated of isolated while he flew by himself in the command module. For forty-five minutes during every lunar orbit, he would lose all contact with Earth, being in the shadow of the dark side of the moon. In his book *Carrying the Fire*, Collins reports his feelings:

> I don't mean to deny a feeling of solitude. It is there, reinforced by the fact that radio contact with the earth abruptly cuts off at the instant I disappear behind the moon. I am alone now, truly alone, and absolutely isolated from any known life. I am it. If a count were taken, the score would be three billion plus two over on the other side of the moon, and one plus God only knows what on this side. I feel this powerfully— not as fear or loneliness—but as awareness, anticipation, satisfaction, confidence, almost exultation. I like the feeling.[18]

In Clark and Graybiel's original break-off study there were reactions reported that ranged from anxiety to elation. Collins wrote that he was something of a loner, so perhaps that steered his reaction upon transiting the far side of the Moon by himself.

LONG-DURATION FLIGHTS

The idea of a human mission to Mars has been around since the dawn of the rocket era. Many aspects of such a mission are daunting: getting the payload to Mars and back safely, minimizing the effects of ionizing radiation from cosmic rays, to name two major concerns. Beyond this is the psychological stress on crew members—both the long duration of the flight and being in relatively cramped quarters with the same crew members. There is also no possibility of a rescue in case something goes wrong.

The duration of the trip can vary, depending on the relative alignments of the Earth and Mars. Estimates for the trip mission are in the range of roughly 800 to 1,000 days, although there is a favorable alignment in 2033 where a mission could be accomplished in 570 days.[19]

Understanding the effects of prolonged periods in space on humans have been a major goal of space station missions since the early 1970s. The first space station was the Soviet Salyut 1, launched in 1971. The death of the three cosmonauts on return to Earth from Salyut 1 was one of the biggest tragedies for the Soviet program. In 1973 the United States launched Skylab. One of the longest-running space stations was the Soviet/Russian Mir, launched in 1987, and deliberately reentered in 2001 over the Pacific Ocean. The world record for the longest continuous stay in space is held by cosmonaut Valeri Polyakov in Mir for 437 days.

The Mir-Shuttle program was a joint US-Russian endeavor in the 1990s where cosmonauts and astronauts would share the Mir space station. Periodic resupplies were accomplished by shuttle missions, Soyuz, and robotic spacecraft. Multiple crises arose on Mir, including a fire, corrosion on cooling tubes, and a collision with a robotic resupply craft that punctured a hole in one of its modules.

Frank Culbertson, who headed the Mir-Shuttle program for NASA, explained a major lesson learned about the duration of flights, "We learned that long-duration flight on a space station is quite a bit different than short flights on the shuttle. The shuttle is an airplane and the station is a ship and it's going to be at sea for a long, long time. It may be a long way from land, but you still got to keep it afloat."[20]

While experiences on space stations can be helpful, there is an additional major factor in a mission to Mars: difficulty of rescue if things go wrong. There are many analogs found in long terrestrial journeys of exploration. The factors that bring stress are:

Increasing distance from rescue in case of emergency
Increasing proximity to unknown or little-understood phenomena
Increasing reliance on a limited, contained environment.
Increasing difficulties communicating with Base
Increasing reliance on a group of companions
Increasing autonomy from Base technological aid
Diminishing resources needed for life and enjoyment of life[21]

Social psychologist Sheryl Bishop and others analyzed some prominent Earth and space expeditions for difficulties associated with these seven factors. This included the famous Apollo 13 expedition, where an explosion on the service module crippled the mission but the module was successfully brought back to Earth. This had six of the seven elements above.

The highest stress-ranked expedition that Dr. Bishop analyzed was the Lady Franklin Bay expedition led by Adolphus Greely. This ticked off all seven of the stressors in Bishop's analysis. It's worth pausing on the Greely Expedition to imagine a possible analogous situation for a stranded human mission to the surface of Mars.

The purpose of the Greely expedition, as it's come to be known, was to establish a northernmost weather observatory on Lady Franklin Bay on Ellesmere Island in the Canadian Arctic, and make astronomical and magnetic observations, in addition to meteorology. The expedition lasted from 1881 to 1884. The plan was to send vessels once a year to resupply the crew manning the remote outpost.

The expedition ship *Proteus* left St. John's, Newfoundland, in July 1881 and landed on the northeastern end of Ellesmere Island in August, dropping off the twenty-five expedition members.

The year 1881 had been anomalously warm, and when the more typical Arctic conditions returned, an 1882 resupply mission couldn't make it through the sea ice that had been navigated the year before. Likewise, 1883 rescue attempts failed. Finally, in 1884 relief came through, but only seven of the original expedition survived.

As the expedition progressed, the marooned men suffered from dwindling supplies. Most of the deaths were from starvation, hypothermia, and drowning. Greely executed a man for theft of food. One man, Joseph Ellison, had amputations of his hands and feet due to frostbite and subsequent gangrene. Ultimately, he fed himself using a spoon tied to the stump of one of his hands. There were reports of cannibalism of the corpses.

After an attempt to procure meager rations from a cache in 1883, Greely wrote in his journal, "We have been lured here to our destruction. We are twenty-four starved men; we have done all we can to help ourselves, and shall ever struggle on, but it drives me almost insane to face the future. It is not the end that affrights anyone, but the road to be traveled to reach that goal. To die is easy; very easy; it is only hard to strive, to endure, to live."[22] While the expedition, at some level, met its goals and provided useful climate data that is instructive in an era of Arctic warming, the price that the crew paid was harrowing.

If a crew on Mars were similarly stranded, a rescue expedition would be very expensive and would require a substantial lead time. The possibility of a stranding and rescue was explored in Andy Weir's novel *The Martian*.

SIMULATED MARS MISSIONS

Space-station flights can focus on the long durations associated with a possible Mars mission, although the current records don't measure up to the amount of time required for even a short expedition to the red planet, reckoned to be over 500 days. To gain insight into crew interactions and the psychological response of the astronauts, both the European Space Agency and NASA carried out and are carrying out simulations of a Mars mission.

Mars500 Project was a joint simulation effort of ESA and Russia, with an isolation chamber designed to mimic both the spacecraft and Mars surface for the expedition. Time delays of up to twenty minutes in communication with the "ground" were built into the simulation.

There was an initial 105-day trial period, followed by a full-on 520-day study that had six crew members: three Russian, two European, and one Chinese. The project concluded in 2011.

Four of the six subjects in the 520-day study had considerable problems with sleep. The crew required more and more sleep over the course of the study, sleeping significantly longer on the "return" flight from the outgoing. One of the major problems in the study was a lack of variety. Matias Basner from the University of Pennsylvania studies the effects of lack of sleep. Basner reported, "The monotony of going to Mars and coming back again is something that will need to be addressed in the future. You don't want your crew hanging around doing nothing."[23] Crew members found that their circadian rhythms went out of synch.

One of the crew members had chronic sleep deprivation that was severe enough to create the majority of errors on a computer test designed to measure alertness and concentration. Another suffered from mild depression. On the other hand, there did not appear to be any interpersonal conflicts.

As of this writing, NASA is in the process of holding its own Mars-simulated missions. This is abbreviated CHAPEA for Crew Health and Performance Exploration Analog. The first of three such studies commenced June 25, 2023, with a four-person volunteer crew. As with the Mars500 study, the crew have several mission-related activities, including simulated spacewalks, exercise, and, importantly, the known stressors of resource limitation, and equipment failure.

None of the four initial volunteers trained as astronauts. Rather, the crew consisted of a research scientist, a structural engineer, an emergency medical physician, and a microbiologist. The four-person crew live in an isolated 1,700-square-foot habitat. Communication delays up to twenty-two minutes are built into the program.

Applicants for the program must be between the ages of thirty and fifty-five, have a master's degree from a STEM field and two years of work toward at doctoral program in related science, engineering, or math field. Finalists undergo medical and psychological testing. The first crew completed their 378-day simulated mission in July 2024. There were no reports of major issues and it seemed to go off well.

It's debatable whether these simulated missions are stringent enough to give insight into what a true Mars expedition would be like. There are the possibilities of multiple equipment crises, like on Mir, but without the possibility of escape or help. Knowing that one is on solid earth in a simulation eliminates what probably is the most major source of anxiety—the knowledge of remoteness. On the other hand, we've seen personalities like Michael Collins who almost revel in such conditions.

It's debatable whether the public will be willing to commit to a mission to Mars by 2033 with the associated costs. Long-term projects like this or committing to mitigating the effects of human-induced climate change are challenging to maintain the general public's attention when so much is focused on fleeting news items.

THE OVERVIEW EFFECT

One curious perspective on the psychological effects of spaceflight was articulated by author Frank White, who wrote *The Overview Effect: Space Exploration and Human Evolution*. By White's account, his inspiration came when he peered out the window of an airplane on a flight across the United States. He felt a profound transformation on seeing the ground from the height of 30,000+ feet and pursued this sense by reading the accounts of astronauts who had a similar transformation. In essence, White suggests that seeing the Earth from space generates a strong shift of perspective for the viewer, where the Earth is seen in a different light—as a tightly linked entity unto itself. Rather than dividing the Earth into continents, countries, cities, oceans, and so on, the new perspective is of how everything is tied together with huge interdependencies, albeit fragile. Some have suggested that the *overview effect* might signal a long-term shift where humanity finally addresses climate change, if only enough people are able to view the Earth from space.

White defines the *overview effect* as:

> The Overview Effect is a cognitive shift in awareness reported by some astronauts and cosmonauts during spaceflight, often while viewing the Earth from orbit, in transit between the Earth and the moon, or from the lunar surface. It refers to the experience of seeing firsthand the reality that the Earth is in space, a tiny, fragile ball of life, "hanging in the void," shielded and nourished by a paper thin atmosphere. The experience often transforms astronauts' perspective on the planet and humanity's place in the universe. Some common aspects of it are a feeling of awe for the planet, a profound understanding of the interconnection of all life, and a renewed sense of responsibility for taking care of the environment.[24]

One of the influences on White's outlook is the Gaia hypothesis. *Gaia* is the ancient Greek name for a primordial goddess of the Earth. The hypothesis, articulated by chemist James Lovelock and biologist Lynn Margulis, states that the Earth is a giant interlocked system where feedback loops that include all living organisms stabilize the climate, the atmosphere, and the oceans to allow for living beings to flourish.

Lovelock suggested that while the atmosphere of Mars seemed to follow the usual laws of thermodynamics, Earth seemed to violate those laws, suggesting that the biological presence of life stabilized our atmosphere in a disequilibrium way. As of this writing, substantial effort is being put into examining the composition of the atmospheres of exoplanets with telescopes such as the James Webb Space Telescope to look for evidence of life outside our solar system.

The fundamental perspective shift that White asserts is that the view of Earth from space directly leads viewers to grasp an intuitive feeling of the interconnectedness associated with the Gaia concept. Moreover, White takes this one step further and asserts a kind of collective consciousness associated to the entire universe that he calls the *Cosma hypothesis*. In many ways, this concept harkens back to rocket pioneer Tsiolkovsky and panpsychism that I mentioned at the chapter's opening.

Perhaps it doesn't bear harping on it, but the concept of a universal collective consciousness is in tension with Liu Cixin's *Dark Forest* model (see chap. 6) where advanced extraterrestrial civilizations are in dire competition with each other and must eliminate competitors. At some level, White's view is a kind of extreme egocentrism with a centrality of human-like civilizations, whereas Liu Cixin's is an allo-centric perspective with secretive civilizations trying to avoid contact.

White imagines a hierarchy of connectivity among progressively larger units. *Terra* represents the earth as a Gaia-like entity. Beyond this is *Solarius*, which represents a kind of integrated, interrelated solar system civilization. Past that is White's concept of a galactic civilization called *Galaxia*.

Creating a Solarius would present a real challenge as none of the other planets seem hospitable to life as we know it. One possible exception to the imagination is Mars, which has often been the subject of much science fiction that describes the possibilities of habitation.

Evolutionary biologist Lynn Margulis, one of the early proponents of Gaia, suggested that Mars could be colonized by introduction of Earth species in what she refers to as "the reproduction of Gaia by budding."[25] The idea of terraforming Mars arises over and over, but it's at best challenging. Mars is on the edge of what we consider a habitable zone, with cold temperatures. Since gravity is weak, the atmosphere is quite thin. With a weak magnetic field, solar wind bombardment also poses a problem. Nonetheless the concept of terraforming Mars to colonize persists and is even the subject of a popular board game.

AMBITIONS AND LIMITS

A human-crewed mission to Mars seems just at our limit with exist-ing technology. The cost is immense and will take a sustained effort and willpower to make it happen. There are also ambitions to crew a permanent base on the Moon, and the concept of mining asteroids for precious metals that are in high demand on Earth. In 2023 a Russian probe accidentally crashed into the Moon, while an Indian probe successfully landed near the Lunar South Pole. The US-led Artemis project aims to return humans to the Moon by the end of the 2020s.

Robotic probes have flown much further than human vessels. For sake of comparison, the Moon is 380,000 kilometers, Mars is about 60 million kilometers away at our point of closest approach. The farthest a satellite has penetrated space is Voyager 1. It was launched in 1977 and is now reckoned to be about 25 billion kilometers away. The distance to the nearest star, Proxima Centauri, is almost 1,600 times further at 40 trillion kilometers.

Voyager 1 is traveling at 17 kilometers/second. At that speed, it would take roughly 75,000 years to reach Proxima Centauri, if it were headed in that direction. If we could travel at the speed of light, and we can't, it would take over four years to reach our nearest neighboring star. Even if we think about approaching the speed of light, the distances and times for interstellar travel are prohibitively large and long. But the human imagination always outruns technical and physical limits. Books, movies, and TV series have looked toward interstellar travel and have developed various fictional schemes for evading the laws of physics that we know of.

The *Star Trek* series popularized a fictional concept of a *warp drive*, where distances in space are shrunk and spacecraft don't violate the speed of light but move through the shrunken space. In a similar vein, the *Star Wars* series invokes a concept called *hyperdrive*, where spaceships travel through the interstellar void in an alternate dimension of *hyperspace*. There's no physical basis to either of these, but they represent clever imaginary schemes that enable the protagonists to visit distant star systems and galaxies on human timescales.

Another fictional schema for crossing the huge voids in space is a wormhole. This at least has more of a known physical basis. Wormholes are possible in Einstein's formulation of general relativity and are a kind of space-time tunnel that can connect very distant parts of space as a short-cut. Two movies made use of wormholes as a plot device for faster-than-light transport. In *Contact*, a movie based on Carl Sagan's novel of the same name, the protagonist, played by Jodie Foster, is transported through an alien-designed spacecraft that traverses a system of wormholes put in place by an ancient cosmic civilization. Another movie that utilized the concept was *Interstellar*, where multiple spacecraft traverse a wormhole that mysteriously appears near the orbit of Saturn. Astronauts used this to scout out planets where we might suitably inhabit, as the Earth was ruined through human-induced climate change.

The concept of wormholes is compelling, much work has been done on understanding what their properties might be, but we have yet to see evidence that they exist, much less whether they can be constructed. The plot twist in *Interstellar* is that wormholes emerge from a quantum theory of gravity.

Will these flights of imagination allow us to slip the known bounds of space-time? It's hard to say one way or the other. In the way Daedalus slipped the bondage of the two-dimensional maze by traveling in a third dimension, perhaps we'll find new wings into another as-yet hidden dimension, and perhaps not suffer the fate of Icarus.

* 14 *

PERSPECTIVES

Throughout, I've written about how striving for new perspectives of space has created social ramifications, which in turn can create new visions of space. Returning to the large particle experiments discussed in chapter 11, we can point to two major social influences associated with the collaborations running them.

Outside of the sheer scale of the experiments in terms of size, complexity, and cost is the social challenge of the collaboration itself. There is the question of coherent access to information among thousands of collaborators; on the other hand, there also is the question of access to data and computing power. Both have parented creative solutions that are now part of the fabric of everyday life.

Consider communication among the collaborators on the LHC experiments. We all need access to technical notes, details of an analysis, even phone numbers and e-mail addresses of people. By the late 1980s the scale of collaborations in particle physics was pushing the boundaries of what could be done with existing Internet practices. It became impossible to send e-mails to hundreds of collaborators without it turning into unwanted spam, but that's where we were. Moreover, collaborators were scattered across a huge number of institutes across the globe, crossing just about every time zone. Access to information was difficult because it was distributed higgledy-piggledy on many computers, not only those at CERN but at the many institutes.

In 1989 then–CERN scientist Tim Berners-Lee, who specialized in computing, proposed a solution: the World Wide Web. In the overview section of Tim's proposal for the Web, he emphasizes the challenges of information in the era of the Large Hadron Collider:

> Many of the discussions of the future at CERN and the LHC era end with the question, "Yes, but how will we ever keep track of such a large project?" This proposal provides an answer to such questions. Firstly, it discusses the problem of information access at CERN. Then, it introduces the idea of linked information systems, and compares them with less flexible ways of finding information.[1]

Tim developed the first web browser and server, and Hypertext Markup Language (HTML). By 1994, further developments of the web led to Mosaic, the first widely available web browser. NCSA's (National Center for Supercomputing Applications) Mosaic was my early workhorse for the web. It was easy to install and use. Later in the 1990s, more web browsers like Netscape Navigator and Internet Explorer became available. In 1994, Yahoo! was launched by Stanford graduates Jerry Yang and David Filo, originally under the name "David and Jerry's Guide to the World Wide Web," with a list of websites. The popularity of the web grew exponentially, leading to the famous dot-com bubble.

Now the web is ubiquitous with, indeed, worldwide uptake. A fascinating characteristic of the modern web is its existence as a new kind of space we inhabit. Even the language of the web employs spatial terms. We may enter a collection of websites through a *portal*. We figure out how to *navigate* around once we cross through the portal. The websites themselves often present themselves as two-dimensional visual maps to their features. Hyperlinks can give access to nearby websites that house important adjacent information. We discuss the merits of various social media platforms. Our likes and dislikes are sorted by AI algorithms that are used to present advertisements for goods that are close to our needs. Political opinions are shown for their proximity to what the algorithms seem to think resonates with us. We might exist in Internet echo chambers, having found like-minded people, perhaps physically far away. Physical adjacency is replaced by a different kind of adjacency.

For me personally, one of the more delicious ironies of the web is that it enabled conspiracy theorists to find each other, including the flat-Earthers we met at this book's start.

Regarding large-scale computing, the experiments operating at CERN generate a huge amount of data, and analysis of that data requires an immense amount of compute power.[2] Just as the problem of information access to large collaborations spread out across the globe, access to data and computing presents a challenge. In the late 1990s the concept of a hierarchy of computing centers arose to serve the needs of the LHC experiments. CERN would be the Tier 0, on top of the hierarchy. Then just below the Tier 0 there would be Tier 1 centers that serve the needs of individual countries. Below these would be Tier 2s that serve regional needs, and finally Tier 3s that serve the needs of users at institutes.

The question became how to link all these computing centers together seamlessly. They may have different operating systems and different forms of data storage. Here is where the concept of *grid computing* came into play. Those of us interested in the problems of computing for the LHC teamed up with computer scientists who were developing grid computing. In effect, we became the testbed for the concept. The name "grid computing" comes from power grids. We may have electricity available from any one of a number of power plants, such as gas, nuclear, and wind turbine. When we plug our toaster into the wall, we don't think about where the power is coming from, nor should we care. In the case of grid computing, we get data and compute power from a computing grid and shouldn't care where the resources originate. The translation from the language of the grid into the operating systems and storage facilities should be seamless, and a key aspect of the grid is *middleware* that straddles between the user and the computer centers.

The computing grid inspired the concept of cloud computing, where our storage and sometimes compute power comes from the cloud. Somehow the name "cloud computing" conjures an ethereal virtual space of data floating around out there, but it has to be housed on true physical servers. In a very real way, the web and cloud computing stem from the challenges we faced for the large experiments on the LHC.

EGOCENTRIC AND ALLOCENTRIC REPRESENTATIONS OF SPACE

Previously, I've employed the concepts of egocentric and allocentric representations to examine new perspectives of space as they emerge. Let me remind the reader of the original definitions of the terms. An *egocentric* representation of space is when landmarks and directions are referenced to the person. An *allocentric* representation of space is when there is a generalized representation of landmarks and directions that doesn't require the positioning of the individual.

What I've tried to do throughout the book is present different perspectives of space, often giving them an explicit label of "egocentric" or "allocentric," but often not. I did not want to give the impression that this is a perfect binary or that it represents some kind of meta-theory of spatial representations, but it does seem like a concept that is extensible to new domains.

I've broadened the definitions to consider a spatial representation egocentric if it necessarily involves humans. This could be in the case where the Earth is the center of the universe. Another example would be if a representation necessarily involves an observer as an integral part. On the other hand, a representation could be considered allocentric if it does not rely on humans. Furthermore, allocentric representations could be taken as a broadening of a perspective of space, like the expansion of the universe.

With these broader definitions of the concepts, although not necessarily precise, we can walk through some of the representations we've seen in the preceding chapters, and where I wasn't explicit, I can be more explicit here.

Most of the models of space from ancient Greece had the Earth as the center of the universe and were hence egocentric in the broadened definition. This would include the modern flat-Earthers. The two departures from this were the models of Philolaus and Aristarchus. The model of Aristarchus had the Earth orbiting the Sun. In the model of Philolaus, the Earth and Sun both orbit the Central Fire.

Astrology puts humans at the center of the affairs of the stars and borrows from Ptolemy's model, so it would seem to be an egocentric representation. Although celestial navigation was first derived from the practice of making natal horoscopes, it seems a bit more allocentric, but at the same time it is necessarily referenced to observations by the navigator, so it still has an egocentric flair to it, even if we see it as a modern, science-based practice.

As Dante's *Divine Comedy* is built explicitly on Aristotle's universe, and explicitly is about humans populating Hades, Purgatory, and the outer spheres of the Moon, Sun, and planets, it would fit into the domain of an egocentric representation. But the landmarks of the *Divine Comedy* exist independent of Dante himself as he navigates Hell, Purgatory, and Heaven. Nonetheless the entire structure is defined by the sins and virtues of human souls.

The development of modern astronomy and physics from Copernicus and Galileo onward would seem to be mostly allocentric, as the representations of space don't rely on the presence of humans, although this is not fully the case, as there have been egocentric temptations along the way. In general, the visions of space have evolved to encompass the larger and the smaller distance scales.

In quantum mechanics there was the notion that consciousness somehow plays a role in the collapse of the wave packet or relies on human measurements. Although that's not strictly the case, the invocation of a conscious observer seems like an egocentric flirtation among some of the advocates of this viewpoint.

The question of extraterrestrials perhaps evades the definitional divide, but there are some curious elements to it. Recall the concept of a plurality of worlds. On the one hand the concept seems to be allocentric because it suggests that we are not alone or unique. On the other hand, the belief in intelligent civilizations in outer space has some elements of projection of human culture, at the very least in the science fiction of extraterrestrials. This has an egocentric flair to it. The rocket pioneer Tsiolkovsky advocated a kind of manifest destiny of humans into outer space, which seems to imply an egocentric projection.

Einstein's special relativity has a fascinating mix to it. Einstein sought a transformation law that would make the speed of light a constant in all relative frames of reference. It also had to break down to the Newtonian/Galilean transformation when the relative velocities were small. This ambition certainly seems to be allocentric in nature but recall that he had an "ah-ha!" moment in his late-night discussion with Michele Besso. Einstein was very much taken by Ernst Mach's philosophy that only relations between observations are real. At a certain point, Einstein pondered observers with watches comparing notes in different reference frames,

> It might appear possible to overcome all the difficulties attending the definition of "time" by substituting "the position of the small hand of my watch" for "time."[3]

By thinking of individuals observing positions on a meter-stick and time with a clock, Einstein was able to bootstrap himself into dropping simultaneity and attaining the transformation law he sought. So, in a sense, he pondered an egocentric viewpoint into a more universal transformation law.

Max Planck had issues with Mach's formulations and made this clear in a rather stark egocentric/allocentric attack, where he wrote of Mach's philosophy, "Nothing is real except the perceptions, and all natural science is ultimately an economic adaptation of our ideas to our perceptions." Planck then articulated his view, that science is "the finding of a fixed world picture independent of the variation of time and people . . . the complete liberation of the physical picture from the individual."[4]

In 1913 Einstein was still a believer in Mach, particularly his conjecture about the influence of fixed stars defining a "zero" of acceleration for the observer that we saw in chapter 8. As a test of his general relativity approached, he wrote to Mach:

> Next year at the solar eclipse it will turn out whether the light rays are bent by the sun, in other words whether the basic and fundamental assumption of the equivalence of acceleration of the reference frame and of the gravitational field really holds. If so, then your inspired investigations into the foundations of mechanics—despite Planck's unjust criticism—will receive a splendid confirmation.[5]

While this clearly expresses admiration, Mach had moved on to disavow Einstein. In the preface to his book, *The Principles of Physical Optics*, published posthumously, Mach writes, "I am compelled in what may be my last opportunity, to cancel my views of the relativity theory."[6] Mach goes on to say that relativity was dogmatic. Many have remarked, including the historian of science Gerald Holton, whom I got the above translations from, that Einstein must have been crushed by Mach's rejection.

The development of models of the universe on progressively larger and smaller scales seems to be the kind of allocentrism that Planck advocates in the quote above, namely that it does not invoke people. Many writers have said that our present models of the cosmos and in the domain of elementary particles is the single most encompassing theory ever. While this may seem to be the triumph of an allocentric perspective, we have the problem of spanning such large distance scales with one theory and making sense of the large number of constants of nature. Since the current theoretical picture points to the possibility of a huge number of universes, each with their own physical constants, there is a strong temptation to invoke the anthropic principle. At least among some physicists, this is their answer. But I submit that the anthropic principle falls into the domain of the egocentric: explaining our universe by invoking our presence.

ONWARD

What the future may hold is anyone's guess, but I suspect that we'll have new cultural manifestations any time a major new vision of space is articulated. There will continue to be a modest tug-of-war between the allocentric and egocentric, although it seems that, in the end, the allocentric usually gets the upper hand. Still, the tension of allocentric and egocentric perspectives played a role in Einstein's development of relativity.

The egocentric/allocentric divide is useful in analyzing concepts of space. Pondering this thought and examining the development of Western physics and astronomy with this binary in mind has been illuminating. At this juncture, I wonder if other extensions of the concept would be helpful to analyze other human endeavors.

Experimentally, we still have gas in the tank. A high-luminosity era at the Large Hadron Collider can probe details of the Higgs boson and open up probes for exotic forms of matter that may help us close in on the fine-tuning problem. Future accelerators that are contemplated can extend the horizon to the smaller and can sharpen these searches and probes of the Higgs. Terrestrial experiments for dark matter continue to evolve. We are entering an era of more precise cosmological probes, both in experiment and detailed theoretical synthesis in astrophysics. Planned experiments on the cosmic background radiation should be able to access the period beyond the 380,000-year blackout and probe aspects of cosmic inflation. The next class of huge telescopes will surely extend our horizon toward the edge of the visible universe, and we're already getting excellent new results from the James Webb Space Telescope.

Fundamental theory seems to be a bit more stuck, and I personally am cheering for a new organizing principle that could lead us forward and away from a reliance on an egocentric explanation for our universe.

And what of our physical exploration? Will we have a human-crewed expedition to Mars? That likely depends on whether the public's imagination can be captured in a sustained way to fund it, along with many issues, both technical and psychological, that must be understood. At some level the sustained public imagination is also a challenge for addressing climate change.

The process of science is arduous. We absolutely need experiments and observations to make progress in physics and astronomy, but their scope and investment of time can be daunting, requiring the stamina of a marathon runner. Nonetheless, the payoffs can be grand as we find our way.

> My guide and I came on that hidden road
> to make our way back into the bright world;
> and with no care for any rest, we climbed—
>
> he first, I following—until I saw,
> through a round opening, some of those things
> of beauty Heaven bears. It was from there
>
> that we emerged, to see—once more—the stars.

(*Inf.* 34 133–39)

ACKNOWLEDGMENTS

Over the years of pondering this book, I've received much input from many sources, too numerous to name, so I will limit myself to a core group to whom I owe the most gratitude.

First to Rick Feinberg, who invited me to the 2012 meeting of the Association for Social Anthropology in Oceania. This provided the initial seed of the approach I took in this book. At this meeting I met Joe Genz, who invited me to the pivotal 2015 voyage in the Marshall Islands that I describe in the preface.

On this project, two Harvard faculty served as an inspiration to me over the years. Owen Gingerich did much historical research on the development of astronomy from Copernicus through Kepler and had an amazing collection of star almanacs from that period. Owen very patiently explained astrology to me and the evolution of thought in that era. Gerald Holton also influenced me through his writings and interactions. I feel blessed to have been in their presence and to have experienced their willingness to share knowledge so readily.

Next, to the Harvard First-Year Seminar program that allowed me to develop ideas that emerged from my trip to the Marshall Islands. Also to the students taking the seminar who pushed me more and more into the realm of applying the new tools I learned to my field of physics and also to astronomy.

I give a major shout-out to my editor, Joe Calamia. I forget how long ago I gave him the initial pitch for this book, but it was sometime in the late 2010s. Joe expressed interest, but we partly lost touch. Joe was persistent and got back in touch with me and helped me develop it into the project it currently is. Moreover, he has been encouraging throughout the writing and editing process. His attention to detail is amazing, and I really must tip my hat to his persistence throughout.

Finally, and ranking up there with my debt of gratitude to Joe, is Cheryl Blanchard. Cheryl took my online course "Backyard Meteorology" and contacted me about it. We had some back-and-forth via e-mail. At a certain point, she said that she was willing to read drafts of chapters for the book. Her comments and cheerleading got me through this process. If a chapter felt flat, she didn't shy away from saying so. If the revised chapter was an improvement, she'd also say so. Her help as a beta reader cannot be overstated.

GLOSSARY

acceleration: The change in velocity over time

action at a distance: Forces like electricity and gravity, where two objects do not physically touch, but the force between them is transmitted through space.

allocentric: In cognitive psychology, representations of space that are generalized and do not place a primacy for the individual or humans. This is an expanded definition for the purposes of this book.

alpha particles (or rays): Particles that produce thick short tracks appearing in a cloud chamber, now known to be helium-4 nuclei (two protons and two neutrons).

anthropic principle: The concept that humans exist because we happen to inhabit a universe where the constants of nature allow for our existence.

antimatter: Spin ½ particles (fermions) that have the same mass, but opposite charge of their matter partners.

ATLAS experiment: A large experiment that records the results of high-energy proton-proton collisions at the LHC at CERN, the sister experiment to CMS.

baryon: A particle that feels the strong, weak, and electromagnetic forces, made of three quarks.

beta decay: The decay of a nucleus into a daughter nucleus plus an electron (or positron) and an antineutrino (or neutrino). This also is responsible for the decay of a neutron to a proton with the same end-products of an electron and an antineutrino.

beta particles (or rays): Particles that produce short spindly tracks appearing in a cloud chamber, now known to be electrons.

Big Bang: A colloquial name for our current model of the universe, characterized by an initial point of infinite density.

Big Bang nucleosynthesis: The burning of hydrogen into heavier elements in the earliest period of the universe.

black body: An idealized object at the same temperature as surrounding electromagnetic radiation.

black-body radiation: The electromagnetic radiation in thermal equilibrium with a black body.

blue shift: The shift of color into shorter wavelengths from celestial object moving toward us.

bosons: Elementary particles with integer spins: 0 or 1, typically.

break-off phenomenon: A feeling of detachment from the Earth by pilots who fly/flew at very high altitudes.

canto: Subdivision of a long poem, like the *Divine Comedy*, from Latin cantus or song.

cathode rays: Particles boiling off an electrically charged plate, now known to be electrons.

causality: The concept that one event can influence another event. The two events must be within reach of light traveling between them.

CERN: Abbreviation for the European Center for Nuclear Physics (from the French Conseil Européen pour la Recherche Nucléaire), a large accelerator facility located outside of Geneva, Switzerland.

celestial equator: Projection of the equator onto the sky

celestial North Pole: Projection of the North Pole onto the sky

celestial South Pole: Projection of the South Pole onto the sky

celestial sphere: Projection of the Earth's coordinates onto the sky

Cepheid (or Cepheid variable): A variable star having a regular cycle of brightness with a frequency related to its luminosity, so allowing estimation of its distance from the earth.

charge screening: The reduction of the charge on particles at larger distances due to surrounding electron-positron bubbles that get polarized.

classical mechanics: The laws of motion and forces as articulated by Galileo, Kepler, and Newton.

cloud chamber: A chamber that images charged particles from a trail of water vapor that condenses along their path.

CDF: Abbreviation for Collider Detector at Fermilab, one of the two detectors to first find the top quark.

CMS: A large experiment that looks at the results of high-energy proton-proton collisions at the LHC at CERN, the sister experiment to ATLAS.

color: A kind of charge associated with the strong interaction. Quarks have three possible kinds of color charges.

confinement: The process of producing color-neutral hadrons from quarks; as a result, no free quarks are seen.

Copenhagen interpretation: A collection of interpretations of quantum mechanics related to the shift from a wave function to a probability of finding a system in a given final state.

Copernican principle: The concept that, on average, the universe will look the same as viewed from any location (for the purposes of this book this is taken to by synonymous with the cosmological principle).

cosmic distance ladder: A way of establishing distances in the universe by using a combination of parallax (close in), Cepheid variables (farther out), and SN1a supernovae.

cosmological principle: The concept that, on average, the universe will look the same as viewed from any location.

cosmic radiation or cosmic microwave background radiation (CMB): Relic electromagnetic radiation in the microwave frequency range, left over from the Big Bang, after the universe cooled to allow light to travel freely.

cosmological constant or Λ: An energy associated with the vacuum that pushes space apart.

dark energy: An energy associated with space itself that is accelerating the expansion of the universe. This is associated with the cosmological constant.

dark matter: A kind of matter that appears to be a large part of the matter of the universe, and yet does not manifest itself as emanating light, such as stars. Dark matter is more widely distributed than ordinary matter in the universe.

deferent: Displacement of the Earth from center of planetary orbit in Ptolemaic model.

Doppler shift: The shift of frequencies in light or sound due to the relative motion of two objects.

ecliptic: Path projected onto the sky of the plane containing planetary, solar, and lunar positions.

egocentric: In cognitive psychology, representations of space represented with respect to a person. In larger terms, for the purposes of this book, I have expanded that to mean representations of space that include humans.

electromagnetic field: A combination of electric and magnetic fields consistent with the rules of the theory uniting electric and magnetic forces. Often, this is used to describe electromagnetic waves that include light and radio signals. The waves are oscillating electric and magnetic fields that travel through space at the speed of light.

electron: Point-like charged particle that is the lightest of the charged leptons.

electroweak force: The force associated with the unification of QED and the weak interaction.

ellipse: An elongated closed two-dimension figure with two foci.

elongation: Angular distance of a planet from the Sun as viewed from Earth.

entorhinal cortex: Part of the brain's limbic system that contains grid cells. It is believed that the interaction of the grid cells in the entorhinal cortex with place cells in the hippocampus generate the cognitive map.

epicycle: Circular motion of a planet in its orbit in Ptolemaic model

equinox: The time of the year when the day and night are of equal duration throughout the Earth.

equivalence principle: The concept that inertial and gravitational mass are equal.

ether: An invisible medium posited to be the carrier of electromagnetic radiation in the late nineteenth century.

felt presence: The perception that there is another person present when there is no person present.

Fermilab: A particle physics accelerator laboratory located in Batavia, Illinois.

fermions: Elementary particles with spin ½.

fine-tuning problem: In the current Standard Model, for a Grand Unified Theory, the fundamental constants of the theory have to cancel to approximately thirty decimal places unless there is some new theory like supersymmetry.

fission (or nuclear fission): The splitting of an atomic nucleus, typically from being struck by a neutron.

frame of reference: Typically, a coordinate system to specify the location of events, including time. There are other definitions beyond physics.

fusion (or nuclear fusion): The process of two nuclei fusing together to form a heavier nucleus.

Galilean relativity: Transformation laws where space and time are treated as separate and time is regarded as absolute for all frames of reference.

gamma rays: X-ray or higher-energy photons that were originally seen as localized puffs in a cloud chamber.

geodesic (or geodesic path): The path an object follows when freely moving in a gravitational field. This includes light.

gluon: The force carrier of the strong interactions (see quantum chromodynamics).

Grand Unified Theory (GUT): A theory that would unify the strong and electroweak forces. A successful GUT has yet to be realized.

gravitational mass: The mass of an object associated with gravitational attraction to another massive object.

hadron: A particle made up of quarks that feels the strong, weak, and electromagnetic interactions. These are typically mesons and baryons, although more exotic hadrons are possible.

heliacal: The rising of a celestial object just before sunrise.

Higgs boson: Particle associated with interactions that give the fundamental particles (quarks, leptons, W, and Z) mass.

hippocampus: Part of the brain's limbic system that is responsible for declarative memory, cognitive maps (place cells), future imagining, and possibly social organization.

houses: Partition of the sky with respect to the location and time of a person's birth in Western astrology.

Hubble constant: A number expressing the relationship between the distance of a galaxy and its redshift, hence recessional velocity.

Hubble radius: The distance at which a galaxy appears to be moving away at the speed of light.

Hubble volume: The volume of the universe inside the Hubble radius.

hypergolic propellant: Rocket fuel consisting of two room-temperature fluids that react when they come in contact and produce thrust.

inertial frame of reference: A frame of reference that is moving at a uniform velocity, as opposed to a non-inertial frame of reference that may be accelerating.

inertial guidance: Use of the sensed motion of a rocket to stabilize and direct its motion.

inertial mass: The mass of an object associated with its resistance to acceleration.

inflation (or cosmic inflation): A model of the very early universe where the expansion of the universe accelerates at an exponential rate, pushing domains of the universe out of causal contact.

jet: In particle physics, it is a collimated spray of hadrons, the remnants of a high-energy quark or gluon

ΛCDM or Lambda Cold Dark Matter: Our current model of the universe that includes dark energy, dark matter, and our own ordinary matter.

laminar flow: For fluids where the paths of molecules are characterized by stable paths.

LHC: Abbreviation for Large Hadron Collider, a circular accelerator that collides protons together at very high energies.

lepton: Spin ½ particles that feel the weak and electromagnetic interactions, but not the strong interaction. This includes particles like the electron and muon, and also neutrinos.

light cone: Regions where events can influence each other, being within reach by light connecting the events.

limbic system: Deep part of the brain that contains a number of specialized areas including the amygdala (emotional memory), hippocampus (declarative memory, place cells), and entorhinal cortex (grid cells).

Limbo: Part of Hell associated with the unbaptized.

Mach's principle: The idea that the universe as a whole defines a zero of acceleration.

MACHO: Abbreviation for "massive compact halo object." A candidate for dark matter that would be small black holes or small massive dark stars.

magnitude: Measure of brightness of a celestial object.

Malebolge: Ditches in lower Hell in the *Divine Comedy*.

Many-Worlds Interpretation of quantum mechanics: The concept that when a measurement is made on a quantum mechanical system, the wave function doesn't collapse, but the universe forks into all the possible final states.

meridian: An arc projected onto the sky from due north to the zenith to due south. All celestial objects rise from the east, reach their highest point at the meridian, then set to the west.

meson: A particle that feels the strong, weak, and electromagnetic forces, made up of a quark and an antiquark.

missing energy: Energy (really momentum) not seen in a detector but inferred from invoking momentum conservation from visible energies.

modality: A division of zodiacal signs into three parts of the seasons: cardinal, fixed, and mutable.

multiverse: A class of ideas that suggest that there are many possible universes, perhaps some with different fundamental constants from our own.

muon: A charged lepton, approximately 200 times the mass of the electron.

naturalness problem: The large-scale difference between the mass of the Higgs and the scale of a Grand Unified Theory creates either the need for exquisite fine-tuning of constants, or some new physics to arise, such as supersymmetry.

nebula: In modern conceptions, a cloud of gas or dust within our galaxy; in earlier perspectives the class included spiral nebulae, now identified as spiral galaxies.

neutrino: A spin ½ particle with no charge and a very light mass. As far as we know there are three species of neutrinos: electron neutrino, muon neutrino, and tau neutrino.

Olbers's paradox: The concept that, in an infinite universe uniformly populated by stars, the night sky would be as bright as the surface of the Sun.

overview effect: A cognitive shift in awareness reported by some astronauts and cosmonauts during spaceflight that gives them a different perspective of Earth as a highly linked system.

Panpsychism: The concept that the universe/reality has a fundamental mind-like component to it.

parallax: The apparent relative motion of objects as seen by a moving observer. The relative motion arises due to the proximity of the object to the observer. For example, driving down a road, you see telephone poles moving faster than a distant hill. This is often used to determine the distance to nearby celestial objects.

parietal cortex (or lobe): In the neocortex of the brain. Among other functions, it integrates different activities, including navigation and proprioception.

parsec: The distance at which the radius of the Earth's orbit subtends one arc-second, approximately 3.2 light-years.

photon: A quantum of electromagnetic energy.

place cells: Single neurons in the hippocampus that fire when a mammal is in a specific position in space.

Planck's constant: A constant (number) expressing the scale at which quantum mechanics becomes important in physics.

Planck scale: The scale at which we believe the quantum mechanical effects of gravity become apparent.

plurality of worlds: The concept that there are other worlds like our own, with intelligent inhabitants.

positron: The antiparticle of an electron.

quanta: Discrete wave packets of energy.

quantum electrodynamics (QED): The theory of electrodynamics that incorporates quantum mechanics, special relativity, and the properties of charged particles. The force carrier of quantum electrodynamics is the photon.

quantum chromodynamics (QCD): The modern theory of the strong interactions, where three color charges replace electrical charge. The force carrier of QCD is the gluon

quantum entanglement: When the relation between particles in quantum states are intertwined, even at a long distance.

quantum mechanics: A branch of physics that describes matter as having both particle and wave characteristics.

quantum tunneling: A process in quantum mechanics where an object can spontaneously shift from one state to another, where it couldn't classically. For example, it can go to a lower-energy potential valley from a higher one, even though there is a barrier in the way.

quark: A spin ½ particle that has charge in units of ⅓ or ⅔ of the electron/proton charge. They feel the strong, weak, and electromagnetic forces. There are six quarks we know of: up, down, charm, strange, top, and bottom. In addition to carrying electrical charge, they also carry color charges.

red shift: The shifting of color into longer wavelengths from a celestial object moving away from us.

relativity: A description linking events happening in two (or more) different frames of reference.

retrograde: Backwards motion of a planet against the fixed background of stars in the sky.

Schrödinger's cat: An example of a system articulated by physicist Erwin Schrödinger involving a cat in a box where it may or may not be killed. It was intended as an absurd limit to the notion that the observation of the cat kills it.

SETI: Abbreviation for search for extraterrestrial intelligence

sidereal: Pertaining to the stars, for example, sidereal motion, sidereal year.

SN1a supernova: The explosion of a white dwarf star that has accreted mass from a companion and becomes unstable.

space-like separations: Events that cannot be in causal contact with each other, as they are farther away than light can connect them.

space-time: In Einstein's relativity time becomes a function of a spatial coordinate when viewed from a moving frame of reference.

special relativity: Transformation laws comparing events in two or more frames of reference where time is a function of position when comparing two or more frames of reference moving at different velocities.

spin: In particle physics, point-like particles possess an intrinsic spin that are quantized in units of Planck's constant. In those units, there can be spin ½ particles (fermions) or spin 0, 1, or 2 particles (bosons).

Standard Model: In particle physics, this is the accepted theory with quark, leptons, the Higgs boson, W boson, Z boson, gluon, and photon and their known interactions. In cosmology it is the Lambda Cold Dark Matter model.

string theory: A theory of fundamental physics that posits that elementary particles are strings rather than point-like objects.

strong force: The short-distance force responsible for holding the atomic nucleus together and creates the interactions between quarks.

sublunary: Belonging to an imperfect earthly realm, as opposed to a more perfect heavenly realm. This comes from the Aristotelian universe, with the Earth lying inside the sphere of the Moon.

superconductivity: The state of an ultra-cold metal where it loses all electrical resistance.

supersymmetry: A theory that posits that for every fundamental particle, there is a partner particle that differs in intrinsic spin by ½. It has the virtue that it would solve the fine-tuning or naturalness problem. It has been suggested that the lightest supersymmetric particle may be the source of dark matter.

Tevatron: A particle accelerator located at Fermilab that collided protons and antiprotons at high energies. It has been decommissioned.

Theory of Everything (TOE): A term for an aspirational theory that explains all the forces and constants of nature.

time-like separations: Events that can be in causal contact with each other. For example, they can be connected by the speed of light or less.

transformation: Rules that describe how to compare events in two reference frames.

tropical: Pertaining to the Sun.

uncertainty principle: A relation that expresses the tradeoff between localizing a particle in physical space and momentum space. This can also be true for localization in energy and time.

unitarity: The concept that the sum of all quantum probabilities of an event must not exceed one.

virtual particle: A particle that interacts between other particles but does not appear in either the initial or final state.

W boson: Charged force carrier associated with the weak force.

wave function: A generalized description of matter in quantum mechanics. The square of the wave function gives the probability of finding a quantum mechanical system in a given state.

wave packet: A combination of waves of different frequencies that give rise to quanta that are neither infinitely spread out in space nor infinitely localized as a point.

wave-packet collapse: When a measurement takes a wave packet (or function) into a single state out of many possible final states.

weak force: The short-distance force associated with beta decay and how stars burn hydrogen into heavier elements.

white dwarf star: The hot dense core of a star remaining after it has used up most of its nuclear fuel.

WIMP: Abbreviation for weakly interacting massive particle. This is a generic candidate for dark matter.

worldline: The path a unique object takes through space-time.

Z boson: Massive neutral force carrier associated with the weak interaction

Zodiac: A zone in the sky along the ecliptic divided into twelve astrological signs.

NOTES

Chapter One

1. H. Eichenbaum, "Time Cells in the Hippocampus: A New Dimension for Mapping Memories," *Nature Reviews Neuroscience* 15 (2014): 732–44.
2. M. Kozhenvnikov webpage, "Allocentric vs. Egocentric Spatial Processing," *Imagery Lab*, n.d., http://www.mariakozhevnikov.com/project-egocentric.html.
3. S. Vann, J. Aggleton, and E. Maguire, "What Does the Retrosplenial Cortex Do?," *Nature Reviews Neuroscience* 10 (2009): 792–803.
4. Lera Boroditsky, "How Language Shapes Thought," *Scientific American* 304, no. 2 (2011).
5. R. Quian Quiroga, L. Reddy, G. Kreiman, C. Koch, and I. Fried, "Invariant Visual Representation by Single Neurons in the Human Brain," *Nature* 435 (2005): 1102–7.
6. M. Schafer and D. Schiller, "Navigating Social Space," *Neuron* 100 (2018): 476–89.
7. Susan Montague, "Space and Person in the Trobriands: The Self as Living and Dead," *Structure and Dynamics* 9, no. 1 (January 2016), https://doi.org/10.5070/SD991031892.
8. G. Holton, *Einstein, History, and Other Passions* (Addison Wesley, 1995), 143.

Chapter Two

1. Dugan Arnett, "The Point of No Turn: For a Dogged Flat-Earther, It's a Lonely New World," *Boston Globe*, November 29, 2017, 1.
2. Dr. Kandiss Taylor "Jesus, Guns, and Babies," podcast with Flat Earth Dave and Matt Long, n.d., https://www.youtube.com/watch?v=03HGxlWIqlE.

3. Dr. Brian Cox, "Science Page," Facebook video, December 18, 2017, https://www.facebook.com/SciencePagecom/videos/395449294224699/.

4. Richard Feinberg, *Polynesian Seafaring and Navigation: Ocean Travel in Anutan Culture and Society* (Kent State University Press, 1988).

5. Quote from Cicero, *Academica*, circa 45 BCE.

6. Archimedes, *The Sand Reckoner*, n.d., https://www.ucolick.org/~laugh /reckoner.shtml.

Chapter Three

1. For a detailed in-depth discussion consult the following website article: Anthony Mallama, "How Bright Are the Planets?," *Sky and Telescope*, May 26, 2020, https://skyandtelescope.org/observing/measuring-planet -magnitudes/#:~:text=Mercury's%20brightness%20depends%20most %20strongly,functions%20of%20Mercury%20and%20Venus.

2. The only minor problem with star-finding apps is that they often go unsupported after some time, so you may have to hunt around. As of this writing the Stellarium app seems to be a good choice: https://stellarium -web.org/. You can consult my previous book, *The Lost Art of Finding Our Way* (Belknap Press of Harvard University Press, 2015), for details on star-finding.

Chapter Four

1. L. Payton and D. Tran, "Moonlight Cycles Synchronize Oyster Behavior," *Biology Letters* 15, no. 1 (2019): 20180299, https://www.ncbi.nlm.nih.gov/pmc /articles/PMC6371907/.

Chapter Five

1. Christian Blauvelt, "Dante and *The Divine Comedy*: He Took Us on a Tour of Hell," *BBC Culture*, June 5, 2018, https://www.bbc.com/culture /article/20180604-dante-and-the-divine-comedy-he-took-us-on-a-tour -of-hell.

2. Not being a poet, I did not try to create a meter or rhyming schema, simply paying attention to the words.

3. Translations by Allen Mandelbaum can be accessed at Digital Dante: Original Research and Ideas, https://digitaldante.columbia.edu/.

4. "John F. Kennedy Quotations: Responsibility, Personal," John F. Kennedy Presidential Library and Museum, https://www.jfklibrary.org/learn/about-jfk /life-of-john-f-kennedy/john-f-kennedy-quotations.

5. Gabriele Vanin, "The Dating of Dante's Voyage in the *Divine Comedy*," *Journal of Astronomical History and Heritage* 26, no. 4 (2023): 903–22.

6. Vanin, "The Dating of Dante's Voyage."

Chapter Six

1. Lucretius, *On the Nature of Things*, trans. and with an introduction by Martin Ferguson Smith, Kindle ed. (Hackett Publishing, 2007), 62–63.
2. Bruno was known for a number of treatises on the art of memory based on the memory palace technique.
3. Steven Dick, *Live on Other Worlds* (Cambridge University Press, 1998), 11.
4. C. Huygens, *Cosmotheoros: Or, Conjectures Concerning the Inhabitants of the Planets*, https://webspace.science.uu.nl/~gento113/huygens/huygens_ct_en_book_1.htm.
5. Huygens, *Cosmotheoros*.
6. Thomas Dobbins and William Sheehan, *Solving the Martian Flares Mystery*, https://alpo-astronomy.org/jbeish/MartianFlaresALPO.pdf.
7. "A Strange Light on Mars," *Nature* 50 (August 2, 1894): 319.
8. William Sheehan and Thomas Dobbins, "The Spokes of Venus: An Illusion Explained," *Journal for the History of Astronomy* 34, pt. 1, no. 114 (2003): 53–63. Others have disputed this claim.
9. F. Drake and D. Sobel, *Is Anyone Out There?* (Delacorte Press, 1992), 62.
10. S. Sheikh et al., "Analysis of the Breakthrough Listen Signal-of-Interest blc1 with a Technosignature Verification Framework," *Nature Astronomy* 5 (2021): 1153–62.
11. Seth Shostak, "LaserSETI," Laser Institute, https://www.seti.org/laserseti.
12. Jennifer Chu, "E.T., We're Home," *MIT News*, November 4, 2018, https://news.mit.edu/2018/laser-attract-alien-astronomers-study-1105.
13. R. Bracewell, "Communications from Superior Galactic Communities," *Nature* 186 (1960): 670–71.
14. Drake and Sobel, *Is Anyone Out There?*, 131–32.
15. Report by NASA UAP Independent Study, 2023, 25, https://science.nasa.gov/uap/.

Chapter Seven

1. Galileo, "Third Letter on Sunspots to Mark Wesler," in *Discoveries and Opinions of Galileo*, trans. with an introduction and notes by Stillman Drake (Doubleday, 1957), 134–35.
2. Owen Gingerich, "The Great Martian Catastrophe and How Kepler Fixed It," *Physics Today* 64, no. 9 (September 2011): 50–54.
3. Gingerich, "The Great Martian Catastrophe," 54.
4. Gerald Holton, "Johannes Kepler's Universe: Its Physics and Metaphysics," *American Journal of Physics* 24 (May 1, 1956): 340–51.
5. Gale Christianson, "Birth of a Masterpiece," Newton's Dark Secrets, *NOVA*, November 2005, https://www.pbs.org/wgbh/nova/newton/principia.html.
6. Marquis de Laplace Pierre-Simon, *A Philosophical Essay on Probabilities*, trans. Frederick Truscott and Frederic Emory (Wiley and Sons, 1902), 4.

7. Cicero, *On Divination*, Loeb Classical Library, trans. William Armistead Falconer, Loeb Classical Library (Harvard University Press, 1923).

Chapter Eight

1. This is held by a number of people, most notably Paul Johnson, who wrote *Modern Times: The World from the Twenties to the Nineties*, rev. ed. (Harper Perennial Modern Classics 2001).
2. Jeroen van Dongen, *Einstein's Unification* (Cambridge University Press, 2010), 38.
3. G. Holton, *Einstein, History, and Other Passions* (Addison Wesley, 1995), 117.
4. "A Mystic Universe," *New York Times*, January 28, 1928, 14.
5. Einstein recalled this conversation in a lecture he gave in Japan in 1922, concerning the origins of relativity theory.
6. Albert Einstein, "On the Electrodynamics of Moving Bodies," *Annalen der Physik* 17 (1905): 891–921.
7. David Hume, "Of the Other Qualities of Our Idea of Space and Time," pt. 2, sec. 3, of *A Treatise of Human Nature*, https://www.gutenberg.org/files /4705/4705-h/4705-h.htm.
8. Many copies are readily available from different publishers.
9. Martin Gardner, *Relativity for the Million* (Macmillan, 1962), 51.
10. Peggy Rosenthal, *Words and Values: Some Leading Words and Where They Lead Us* (Hamilton Books, Rowman and Littlefield Publishing, 1984), 119–20.
11. Paul Johnson, *Modern Times: The World from the Twenties to the Nineties*, rev. ed. (Harper Perennial Modern Classics 2001), 3.
12. H. Wildon Carr, "Metaphysics and Materialism," *Nature* 108 (October 20, 1921): 247–48.
13. Jose Ortega y Gasset, *The Modern Theme*, trans. James Cleugh (Harper Torchbooks, 1961), 136.
14. Jacques Derrida, quoted by Alan Sokal in "Hermeneutics of Classical General Relativity," his classic "hoax." https://physics.nyu.edu/sokal/transgress_v2 /node2.html.
15. Karl Anderson, "How America Lost Its Mind," *The Atlantic*, September 2017, https://www.theatlantic.com/magazine/archive/2017/09/how-america -lost-its-mind/534231/.
16. David Ernst, "Donald Trump Is the First President to Turn Postmodernism Against Itself," *The Federalist*, January 23, 2017, https://thefederalist.com/2017 /01/23/donald-trump-first-president-turn-postmodernism/.
17. Sean Coughlan, "What Does Post-Truth Mean for a Philosopher?" *BBC News*, January 12, 2017, http://www.bbc.com/news/education-38557838.
18. This argument about a person spinning is often attributed to Steven Weinberg, *Gravitation and Cosmology: Principles and Applications of the General Theory of Relativity* (John Wiley and Sons, 1972), 17.

Chapter Nine

1. Some caution is needed for the curious reader who may wish to investigate *The Sand Reckoner*. The term "universe" is sometimes used to denote the Earth–Sun–Moon system and sometimes used for the sphere of the fixed stars. Care should be exercised to see which Archimedes is referring to.

2. NASA Hubble Mission Team, "Hubble Views the Star That Changed the Universe," NASA, May 27, 2011, https://science.nasa.gov/missions/hubble /hubble-views-the-star-that-changed-the-universe/.

3. Edwin Hubble, *The Realm of the Nebulae* (Yale University Press, 1982), 105.

4. Adam Frank and Marcelo Gleiser, "The Story of Our Universe May Be Starting to Unravel," *New York Times*, September 2, 2023, https://www.nytimes.com /2023/09/02/opinion/cosmology-crisis-webb-telescope.html.

5. Steven Weinberg, *The First Three Minutes* (Bantam Books, 1977), 154.

Chapter Ten

1. Craig Callendar, "Nothing to See Here: Demoting the Uncertainty Principle," *New York Times*, July 21, 2013.

2. Jim Holt, "Uncertainty About the Uncertainty Principle," *Slate*, March 6, 2002.

3. Robert P. Crease, "Too Confident About Uncertainty," *Physics World*, December 2001, 18.

4. There are multiple quotes of Einstein to this effect.

5. Charles Darwin, quoted in "Recollections from the First Copenhagen Conference," in *Selected Papers of Leon Rosenfeld*, ed. Robert Cohen and J. J. Stachel (Springer Verlag, 1979), 310.

6. Fritjof Capra, *The Tao of Physics: An Exploration of the Parallels Between Modern Physics and Eastern Mysticism*, 5th ed. (Shambhala Publications, 2010).

7. David Mermin, "What's Wrong with This Pillow," *Physics Today*, April 1989, 9.

8. Peggy Landsman, "Schrödinger's Cat," originally published in *Scientific American* 326, no. 4 (2022): 24.

9. Jesse Emspak, "Quantum Entanglement: Love on a Subatomic Scale," Space. com, February 14, 2016.

10. Alexander Wendt, interviewed by Cathy Becker, Mershon Center for International Studies, Ohio State University, 2015, https://mershoncenter.osu.edu /news/q-and-alexander-wendt-quantum-mind-and-social-science.

11. Marilynne Robinson, "Humanism, Science, and the Radical Expansion of the Possible," *The Nation*, October 22, 2015, https://www.thenation.com/article /archive/humanism-science-and-the-radical-expansion-of-the-possible/.

12. Dr. Laura Berman, "Quantum Love: Use Your Body's Atomic Energy to Create the Relationship You Desire," 2024, https://drlauraberman.com/product /quantum-love-use-your-bodys-atomic-energy-to-create-the-relationship -you-desire.

13. Barbara Ehrenreich, *Smile or Die*, RSA lecture, January 11, 2010, https://www .youtube.com/watch?v=PJGMFu74a7o.

14. Peter Byrne, *The Many Worlds of Hugh Everett III* (Oxford University Press, 2012), 142.

15. Sean Carroll, "Are Many Worlds and the Multiverse the Same Idea?," *Sean Carroll* (blog), https://www.preposterousuniverse.com/blog/2011/05/26/are-many-worlds-and-the-multiverse-the-same-idea/.

16. As a postscript, Hugh Everett left academia after completing his dissertation and went to work for the Defense Department. He died of a sudden heart attack in 1982, at age fifty-one. His daughter Elizabeth suffered from what appeared to be depression and drug addiction. She committed suicide in 1996. According to Byrne, she left a note: "Funeral requests: I prefer no church stuff. Please burn me and DON'T FILE ME. Please sprinkle me in some nice body of water or the garbage, maybe that way I'll end up in the correct parallel universe to meet up with Daddy" (Byrne, *The Many Worlds of Hugh Everett III*, 342).

Chapter Eleven

1. Hans Kristensen, Matt Korda, Eliana Johns, Mackenzie Knight, and Kate Kohn, "Status of World Nuclear Forces," Federation of American Scientists, March 29, 2024, https://fas.org/initiative/status-world-nuclear-forces/.

2. Author's note: I was asked to blog about the discovery in context for July 5, 2012, but found that just about every science journalist on the planet had already written about it. Personally, I found it a bit annoying that the journalists mostly interviewed theorists and not experimentalists about the discovery. In the end, I wrote a snarky blog post where I imagined the reincarnation of Mark Twain covering the discovery. Huth, "The Celebrated God Particle, by Mark Twain," *Quantum Diaries*, July 5, 2012, https://www.quantumdiaries.org/2012/07/05/the-celebrated-god-particle-by-mark-twain/.

Chapter Twelve

1. Ethan Siegel, "Could the Large Hadron Collider Make an Earth-Killing Black Hole?," *Forbes*, March 11, 2016, https://www.forbes.com/sites/startswithabang/2016/03/11/could-the-lhc-make-an-earth-killing-black-hole/.

2. Dennis Overbye, "Gauging a Collider's Odds of Creating a Black Hole," *New York Times*, April 15, 2008, https://www.nytimes.com/2008/04/15/science/15risk.html.

3. Stephanie Pappas, "German Court Rules that Collider Won't Destroy Earth," *NBC News*, October 19, 2012, https://www.nbcnews.com/id/wbna49483243.

4. Michelangelo Mangano, personal communication via e-mail on his recollection of the conversation.

5. Rowan Hooper, "Multiverse Me: Should I Care About My Other Selves?," *New Scientist*, September 24, 2014.

6. Sam Kris, "The Multiverse Idea Is Rotting Culture," *The Atlantic*, August 29, 2016.

7. James Peebles, "The Physicists Philosophy of Physics," January 29, 2024, https://arxiv.org/abs/2401.16506.

8. Peebles, "The Physicists Philosophy of Physics."

9. Leonard Susskind, interviewed on the PBS series *Closer to the Truth*, March 3, 2020, transcribed by the author.

10. Paul Steinhardt, interview by David Dierter, June 4, 18, and 30, and July 8, 2020, Neils Bohr Library and Archives, American Institute of Physics, College Park, MD, https://repository.aip.org/steinhardt-paul-2020-june-4-june-18-june-30-july-8.

11. Neil Turok and Latham Boyle, "Gravitational Entropy and the Flatness, Homogeneity and Isotropy Puzzle," January 2022, https://arxiv.org/abs/2201.07279.

12. Charlie Wood, "Why This Universe? A New Calculation Suggests Our Cosmos Is Typical," *Quanta Magazine*, November 17, 2022, https://www.quantamagazine.org/why-this-universe-new-calculation-suggests-our-cosmos-is-typical-20221117/.

13. Alexandre Alves, Alex Dias, and Roberto da Silva, "Maximum Entropy Principle and the Higgs Boson Mass," August 4, 2014, https://arxiv.org/abs/1408.0827.

Chapter Thirteen

1. "A Severe Strain on Credulity," Topics of the Times, *New York Times*, January 13, 1920, 12, col 5, accessed from https://web.archive.org/web/20070217065558/http://it.is.rice.edu/~rickr/goddard.editorial.html.

2. This transfer was part of a larger Operation Paperclip that secretly brought a large number of German scientists, engineers and their families to the United States to tap their expertise.

3. President Dwight Eisenhower, Address to the Nation on Future of US Security, Municipal Auditorium, Oklahoma City, Oklahoma, November 13, 1957, American Rhetoric: Online Speech Bank, https://www.americanrhetoric.com/speeches/dwighteisenhowernationalsecurityfuture.htm.

4. John F. Kennedy, in a speech to Congress, May 26, 1961, https://www.nasa.gov/history/the-decision-to-go-to-the-moon/.

5. Brant Clark and Ashton Graybiel, "The Break-Off Phenomenon: A Feeling of Separation from the Earth Experienced by Pilots at High Altitude," *Journal of Aviation Medicine* 28, no. 2 (April 1957): 121–26, 122.

6. Clark and Graybiel, "The Break-Off Phenomenon."

7. David Simons, "The Breakoff Phenomenon During Balloon Flight in the Stratosphere," in *Environmental Effects on Consciousness*, ed. Karl Schaefer (Macmillan, 1958), 91.

8. David Simons, *Manhigh: An Account of a Balloon Flight Into Space* (Doubleday, 1960), 257.

9. Martin Giffen, "Break Off: A Phase of Spatial Disorientation," *US Armed Forces Medical Journal* 10, pt. 2 (1959): 1301.

10. Patricia Santy, *Choosing the Right Stuff: The Psychological Selection of Astronauts and Cosmonauts* (Praeger, 1994), 39.

11. T. S. Eliot, *The Waste Land* (1922); see, for example, https://www.rsgs.org/blog/sir-ernest-shackleton-and-t-s-eliots-third-man.

12. Margaret Weitekamp, *Right Stuff, Wrong Sex* (Johns Hopkins University Press, 2005), 3.

13. NASA, "Project Mercury Overview—Astronaut Selection," November 30, 2006, last updated 2017, https://www.nasa.gov/mission_pages/mercury/missions/astronaut.html.

14. Albert Harrison and Edna Fielder, "Behavioral Health," in *Psychology of Space Exploration: Contemporary Research in Historical Perspective*, ed. Douglas Vakoch, NASA History Series (NASA, 2011).

15. Santy, *Choosing the Right Stuff*, 28.

16. Santy, *Choosing the Right Stuff*, xvi.

17. Santy, *Choosing the Right Stuff*, 30.

18. Michael Collins, *Carrying the Fire: An Astronaut's Journey* (Farrar, Straus and Giroux, 1972), 402.

19. Jamie Carter, "NASA Should Send $17 Billion Human Mission to Mars in 2033, Say Experts," *Forbes*, May 18 2023, https://www.forbes.com/sites/jamiecartereurope/2023/05/18/nasa-should-send-17-billion-human-mission-to-mars-in-2033-say-experts/.

20. Bryan Burrough, *Dragonfly: An Epic Adventure of Survival in Outer Space* (HarperCollins, 1998), 513.

21. Sheryl L. Bishop, "From Earth Analogs to Space: Getting There from Here," in *Psychology of Space Exploration*, ed. Vakoch, 65–66.

22. "The Greely Expedition," *American Experience*, PBS, February 2019, https://www.pbs.org/wgbh/americanexperience/films/greely/#part01.

23. Ian Sample, "Fake Mission to Mars Leaves Astronauts Spaced Out," *The Guardian*, January 2013 https://www.theguardian.com/science/2013/jan/07/fake-mission-mars-astronauts-spaced-out.

24. Frank White, *The Overview Effect*, 3rd ed. (American Institute of Aeronautics and Astronautics, 2014), 2.

25. L. Marguils and O. West, "Gaia and the Colonization of Mars," *GSA Today* 11 (1993): 277–80.

Chapter Fourteen

1. The original proposal, appropriately available on the web: https://www.w3.org/History/1989/proposal.html.

2. I hesitate to give numbers because data storage and computing capabilities are constantly changing, so quoting number as of this writing may be obsolete by the time these words are read.

3. Albert Einstein, "On the Electrodynamics of Moving Bodies," *Annalen der Physik* 17 (1905): 891–921.

4. Gerald Holton, *Thematic Origins of Scientific Thought: Kepler to Einstein* (Harvard University Press, 1988), 245.

5. Holton, *Thematic Origins of Scientific Thought*, 246.

6. Holton, *Thematic Origins of Scientific Thought*, 248.

INDEX

Page numbers in italics refer to figures and tables.

Sputnik, 299–300
Standard Model of particle physics,
     169, *252*, 257, 273; free parameters
     and, 275–76; remaining mysteries,
     275–76, 277
Stapp, John, 301
stars: along the ecliptic, 53; appear-
     ance depending on latitude, 31; of
     Aristotle's Firmament, *37*, 38, 39;
     in Dante's *Inferno*, 79, 82–83, 84,
     87–88, 328; in Dante's *Paradiso*, 79,
     88, 99; in Dante's *Purgatorio*, 79, 88,
     90–91; Earth's axis affecting posi-
     tion of, 59; first formation of, 207;
     heavy elements produced by, 206; in
     Marshall Islands navigation, ix, xi;
     powered by nuclear forces, 237–38;
     twinkling of, 54
Steinhardt, Paul, 291
stelle (stars), 79, 88, 96
stick charts, x, xii–xiii
Strassmann, Fritz, 238
string landscapes, 286, 290
string theory, *284*, 284–86
strong force, 237–38, 249–53, 269
sublunary Earth, 40
Sun: bending of starlight around, 185;
     Dante on path of, 91–95, *92, 94, 95*;
     declination of, 51–52, *52*; influence on
     Earth, 57; in Ptolemy's model, 42–43;
     relative motion against the stars, 26,
     29, 40, 50; rising in east and setting in
     west, 22, *25*, 25–26, 27–29, *29*, 33; tilt
     of Earth's axis and, 50–51, *51*
Sun signs, 60–62, *61, 62*
superconductivity, 271
superior planets, 43
supernovas, 191–92; heavy elements
     produced by, 206
superstition, 20, 21
supersymmetry, 204, 281
survey knowledge, 4–6, *7*
Susskind, Leonard, 290–91

tau lepton, 251, 259
Taylor, Kandiss, 23

technicolor, 272
telescope: extraterrestrial beings and,
     100, 102, 103, 109; of Galileo, 48,
     102; new and planned, 292; planets
     beyond Saturn and, 65; revealing
     structure on the planets, 97, 99, 105.
     See also radio telescopes
Teller, Edward, 239
Tereshkova, Valentina, 300, 307
Tetrabiblos (Ptolemy), 57
Tevatron, 260–62, 265, 268
Thales, 28, 82
thermodynamics, 212, 316
Thomas Aquinas, 39
Thompson, William (Lord Kelvin),
     237
Thomson, J. J., 234–35, 251
three-body problem, 143–44
time: absolute, xiii, 129, 146, 147, 153,
     165, 175; arrow of, 170, 171–72,
     242–43; in *Divine Comedy*, 74, 80–81,
     82–83, 84, 85–86, 87, 88, 90, 91–94;
     frames of reference and, 128; hippo-
     campus in perception of, 10–11;
     language of motion through, 1, 11;
     slower in gravitational field, 183, 185;
     slower in moving reference frame,
     156. *See also* space-time
time cells, 10–11
time-like separation, 170–71, *171*
Tolman, Edward, 4, 5–6
T-O map, 75, *76*
top quark, 259–63, *261, 262*,
     267–68
Torres, Jason, 23
train paradox, *166*, 166–68, *167, 168*
transformation between frames of
     reference, 128, 150–51, 158; Gali-
     lean, 153, 157, 163; observers not
     necessary for, 177–78; preserving
     speed of light, 160–61; of special
     relativity, 163, 165. *See also* frames
     of reference
Trojan Horse, 86
Truman, Harry S., 238–39
Trump, Donald, 180